**Springer Proceedings in Mathematics & Statistics**

Volume 147

More information about this series at http://www.springer.com/series/10533

# Springer Proceedings in Mathematics & Statistics

This book series features volumes composed of select contributions from workshops and conferences in all areas of current research in mathematics and statistics, including operation research and optimization. In addition to an overall evaluation of the interest, scientific quality, and timeliness of each proposal at the hands of the publisher, individual contributions are all refereed to the high quality standards of leading journals in the field. Thus, this series provides the research community with well-edited, authoritative reports on developments in the most exciting areas of mathematical and statistical research today.

Boris Defourny • Tamás Terlaky
Editors

# Modeling and Optimization: Theory and Applications

MOPTA, Bethlehem, PA, USA, August 2014
Selected Contributions

Springer

*Editors*
Boris Defourny
Department of Industrial & Systems Engineering
Lehigh University
Bethlehem, PA, USA

Tamás Terlaky
Department of Industrial & Systems Engineering
Lehigh University
Bethlehem, PA, USA

ISSN 2194-1009 ISSN 2194-1017 (electronic)
ISBN 978-3-319-23698-8 ISBN 978-3-319-23699-5 (eBook)
DOI 10.1007/978-3-319-23699-5

Library of Congress Control Number: 2015953775

Mathematics Subject Classification (2010): 49-06, 49Mxx, 65Kxx, 90-06, 90Bxx, 90Cxx

Springer Cham Heidelberg New York Dordrecht London


Printed on acid-free paper

Springer International Publishing AG Switzerland is part of Springer Science+Business Media (www.springer.com)

# Preface

This volume contains a selection of papers that were presented at the Modeling and Optimization: Theory and Applications (MOPTA) Conference held at Lehigh University in Bethlehem, Pennsylvania, USA, between August 13 and August 15, 2014. MOPTA 2014 aimed to bring together a diverse group of researchers and practitioners, working on both theoretical and practical aspects of continuous or discrete optimization. The goal was to host presentations on the exciting developments in different areas of optimization and at the same time provide a setting for close interaction among the participants. The topics covered at MOPTA 2014 varied from algorithms for solving convex, combinatorial, nonlinear, and global optimization problems and addressed the application of optimization techniques in finance, electricity systems, healthcare, and other important fields. The five papers contained in this volume represent a sample of these topics and applications and illustrate the broad diversity of ideas discussed at the conference.

The first part of the name MOPTA highlights the role that modeling plays in the solution of an optimization problem, and indeed, the first two papers in this volume illustrate the benefits of effective modeling techniques.

The paper by Ilya O. Ryzhov proposes a variety of ways in which Bayesian inference can be integrated to optimization problems under uncertainty where decision makers have the opportunity to learn more about parameters and relations among variables. Approximations are proposed to represent posterior beliefs, which do not admit closed-form descriptions. These techniques broaden the applicability of Bayesian inference, which embodies the key principles of reasoning under uncertainty.

The paper by Miguel Anjos discusses problems involved in the management of the power grid in the presence of distributed generation and renewables. It proposes complexity reduction techniques for the combinatorial search in the presence of symmetric entities, which is a situation to be found in many man-made interconnected systems.

The next three papers in the volume address the other foci of MOPTA, namely optimization algorithms, theory, and applications.

The paper by Hongbo Dong and Nathan Krislock considers the solution of mixed-integer quadratically constrained optimization problems (MIQCP) using relaxations based on semidefinite programming (SDP) techniques. The authors show that the SDP relaxations can be solved approximately using a variety of techniques. MIQCPs are ubiquitous in certain areas of engineering such as process systems, as well as in the general context of combinatorial optimization. Thus, this paper contributes to the development of specific algorithmic techniques for this very important class of problems.

The paper by Sunil Chopra, Sangho Shim, and Daniel E. Steffy investigates the Master Knapsack Polytope, which is an object of fundamental theoretical interest in integer optimization. They identify relaxations that can speed up branch-and-cut algorithms used to solve mixed-integer linear optimization problems. Reducing solution times is desirable in virtually all applications of mixed-integer linear optimization.

The paper by Philip E. Gill, Michael A. Saunders, and Elizabeth Wong considers the solution approach for large-scale nonlinear optimization based on solving a sequence of approximate quadratic models for the Lagrangian, known as sequential quadratic optimization (SQP). Nonlinear optimization is instrumental in many areas of engineering where the exact relation between variables should be preserved. The authors develop convexifications in a variety of ways, which allows them to exploit second-derivative information when it is available.

We thank the sponsors of MOPTA 2014, namely AIMMS, SAS, Gurobi, and SIAM. We also thank the host, Lehigh University, as well as the rest of the organizing committee: Frank Curtis, Luis Zuluaga, Larry Snyder, Ted Ralphs, Katya Scheinberg, Robert Storer, Aurélie Thiele, and Eugene Perevalov.

Bethlehem, PA, USA
June 2015

Boris Defourny
Tamás Terlaky

# Contents

# Contributors

**Miguel F. Anjos**
Canada Research Chair, Trottier Institute for Energy, GERAD & Polytechnique Montreal, Montreal, QC, Canada

**Sunil Chopra**
Kellogg Graduate School of Management, Northwestern University, Leverone Hall, Evanston, IL, USA

**Hongbo Dong**
Department of Mathematics, Washington State University, Pullman, WA, USA

**Philip E. Gill**
Department of Mathematics, UC San Diego, La Jolla, CA, USA

**Nathan Krislock**
Department of Mathematical Sciences, Northern Illinois University, DeKalb, IL, USA

**Ilya O. Ryzhov**
Robert H. Smith School of Business, University of Maryland, College Park, MD, USA

**Michael A. Saunders**
Systems Optimization Laboratory, Department of Management Science and Engineering, Stanford University, Stanford, CA, USA

**Sangho Shim**
Kellogg Graduate School of Management, Northwestern University, Leverone Hall, Evanston, IL, USA

**Daniel E. Steffy**
Department of Mathematics and Statistics, Oakland University, Rochester, MI, USA

**Elizabeth Wong**
Department of Mathematics, UC San Diego, La Jolla, CA, USA

# Approximate Bayesian inference for simulation and optimization

Ilya O. Ryzhov

**Abstract** We present an overview of approximate Bayesian methods for sequential learning in problems where conjugate Bayesian priors are unsuitable or unavailable. Such problems have numerous applications in simulation optimization, revenue management, e-commerce, and the design of competitive events. We discuss two important computational strategies for learning in such applications, and illustrate each strategy with multiple examples from the recent literature. We also briefly describe conjugate Bayesian models for comparison, and remark on the theoretical challenges of approximate models.

**Keywords** Optimal learning • Stochastic optimization • Bayesian statistics • Approximate Bayesian inference

**MSC (2010):** 62F15 Bayesian inference, 62F07 Ranking and selection, 62L12 Sequential estimation

## 1 Introduction

We consider statistical learning problems in which information is collected sequentially. We are specifically interested in the challenges that arise when this process occurs inside multi-stage stochastic optimization problems, in which decisions are subject to uncertainty about the decision-maker's environment. This type of uncertainty is distinct from the usual framework of stochastic programming, in that the probability distributions driving the uncertainty may themselves be unknown. Such problems arise within revenue management, energy, health care, marketing, and a variety of other applications, and may be classified under the broad name of *optimal learning* [29]. In these problems, a single decision may fulfill some economic objective (such as earning revenue), but it also leads to the acquisition

I.O. Ryzhov (✉)
Robert H. Smith School of Business, University of Maryland, College Park, MD, USA
e-mail: iryzhov@rhsmith.umd.edu



B. Defourny, T. Terlaky (eds.), *Modeling and Optimization: Theory and Applications*, Springer Proceedings in Mathematics & Statistics 147,
DOI 10.1007/978-3-319-23699-5_1

of new information about the environment which may improve future decisions (for example, the mean sales observed over a fixed period of time may be used to obtain more accurate information about the demand distribution).

The interplay between statistical and optimal learning motivates the development of sophisticated algorithms that quantify the tradeoff between economic objectives and information. While we briefly describe such algorithms for motivation, our focus in this chapter will be on the statistical side, rather than the optimization side. In the following, we discuss approximate Bayesian inference, a methodological tool that can be used (and has been used, in the literature as well as in practice) to create powerful computational learning models for problems that would otherwise be intractable. Such models can then be combined with various types of optimization algorithms developed in the optimal learning literature.

Bayesian statistics offers an attractive way of modeling environmental uncertainty in optimal learning. The Bayesian philosophy is to model any unknown quantity as a random variable, whose distribution represents the decision-maker's *belief* about the range and likelihood of possible values for the quantity. Probability distributions thus play a dual role in Bayesian statistics. First, we collect random information from the field (e.g., observations of demand). The distribution of these observations refers to exogenous uncertainty (e.g., variation in customer behavior), and may have one or more unknown parameters. These unknown parameters themselves have a probability distribution, which is completely specified by the decision-maker, and represents uncertain beliefs rather than stochasticity occurring in nature.

As more information is collected, the belief distribution is changed (or "updated") by conditioning on all the observations that have been made up to this point. The updating process is particularly simple for Bayesian models that possess the property of *conjugacy*: as long as both the belief distribution and the exogenous noise distribution come from a certain combination of families (e.g., normal and normal, or gamma and exponential), the updated beliefs will belong to the same family as the initial beliefs. Thus, updating reduces to a simple recursive calculation for the parameters of the distribution (e.g., the mean and variance, in the normal setting). It is no longer necessary to store an entire probability density in memory; our beliefs can be completely characterized by a small number of parameters.

This computational convenience may not be of high importance in traditional statistics, where we assume that we are given a dataset from the beginning, with no control over how the observations were chosen. However, conjugacy becomes very important in sequential learning, where observations arrive one at a time and are influenced by our decisions (or by decisions made by some optimization algorithm). For example, if we have observed a month's worth of sales, we would naturally take this information into account when deciding the next day's price; however, that price will in turn influence future observed sales. The algorithms used to make decisions (such as setting prices) rely on the ability to compactly model uncertain beliefs with a handful of parameters, otherwise their computational cost may be too great.

Unfortunately, there are many applications where conjugate models are simply not available, and there is no clear way to force the problem into one of the standard conjugate settings. This chapter discusses approximate Bayesian inference as a possible solution to this problem. If we wish to use the benefits of conjugacy, but no appropriate conjugate model exists, we simply force conjugacy by replacing the updated distribution (typically a computationally "difficult" mixture of some sort) by one from the desired family. The parameters of this artificial distribution can be chosen to optimally approximate the difficult distribution. We consider two techniques for creating such approximations, namely moment-matching and density filtering. These techniques are based on fairly simple principles, but the technical details are different for each application, and can easily become complicated.

The main goal of this chapter is to argue that approximate Bayesian inference is a flexible and useful technique, offering practical solutions even when the precise theory behind the approach is not well understood. We focus our presentation around several applications that have received attention in the recent literature. We will generally not devote much space to the technical details or derivations (references are provided for interested readers), instead focusing on the problems motivating the approximate models, and the final results. We will first give a brief overview of conjugate models (mostly focusing on normal distributions) in Section 3, to provide rigorous definitions for the concepts and to illustrate the advantages of Bayesian updating. In Section 4, we briefly discuss Bayesian inference and conjugacy in optimal learning, using a particular class of simulation problems for context. Then, in Section 5, we briefly describe each approximate methodology and illustrate it with applications. Section 6 briefly describes the theoretical challenges of approximate Bayesian inference and outlines some open problems in this area. Section 7 concludes.

## 2 Applications

We describe several applications where approximate Bayesian inference is relevant. We will return to most of these examples in Section 5 with more formal definitions and exposition; for now, we keep the discussion at a high level to emphasize that the technique applies to a wide variety of important problems.

*Wind farm placement.* Suppose that there is a finite number of candidate locations where we could build a new wind farm. We would like to find the location where the average power output would be the highest. However, this depends on uncertain factors such as wind speed, as well as on complex physical factors such as topography. Practitioners use expensive Monte Carlo simulations [23] to estimate the effectiveness of a wind farm at a certain location. However, our budget for running simulations may be small, while the number of candidate locations may be large. Ideally, we would exploit "similarities" between locations: for example, if two locations are close together and have similar topographical features, we may expect them to perform similarly. Thus, running a simulation for the first location should also provide some information about the second, without requiring us to expend any additional simulations.

This problem can be viewed as an example of *ranking and selection*, a widely used model in the simulation literature (discussed in Section 4). The difficulty in this particular example, however, is that the similarities between alternatives (modeled as correlations between simulation output at different locations) may be difficult to quantify. Essentially, we have to learn them just as we learn the locations' performance values, through the outcomes of individual simulations. Standard statistical models do not provide a way to easily learn unknown correlations from scalar observations; however, approximate Bayesian inference can give us this ability [30].

*Market design.* Consider a market where traders buy and sell an asset. If the traders are well-informed about the value of the asset, we may expect the market to achieve an equilibrium. Suppose, however, that the market experiences a shock, and the value of the asset changes. This now leads to a learning period where traders change their beliefs about the value based on the observed outcomes of their trading strategies. Moreover, many traders may choose to refrain from participating in the market due to increased risk following from the shock. This leads to lower liquidity of the market, which may in turn cause other traders to leave. In such situations, an entity known as a "market-maker" may be employed to trade with other participants [9]. In some cases, all traders may be required to conduct their trades through the market-maker, who may serve as a buyer or seller in a transaction. The market-maker sets prices in an effort to stimulate market activity, and may also seek to maximize revenue from trading. Since the market-maker participates in a much larger number of transactions than any individual trader, it also collects more information about the new value of the asset. We thus need to model the market-maker's learning process in order to develop effective trading strategies. Information should be processed quickly, due to the high volume of transactions.

*Dynamic pricing.* A seller interacts with a sequence of homogeneous customers by offering prices for a product. The seller may use some demand model representing the customers' willingness to pay (for example, a linear regression model for expected sales, or a logistic curve for the probability of a successful sale). The difficulty, however, is that a customer's exact valuation of the product is never observed. The seller only sees whether the customer accepted or rejected a given price. It is then necessary to learn the customer valuations (i.e., to improve an existing demand model) based purely on these censored observations; furthermore, in settings such as e-commerce, it is necessary to adapt to customer behavior as quickly as possible [5]. Using approximate Bayesian inference, one can develop computationally efficient learning schemes.

*Competitive online gaming.* In multiplayer online games [33], thousands of players simultaneously log on to a server and state their willingness to play. Players are not willing to wait long, and so the game master must match each player with an opponent as quickly as possible. However, the quality of each match is also important, as players do not enjoy games where one side substantially outclasses the other. Thus, the game master should ensure that each player is matched to an opponent possessing a similar skill level. In practice, the "skill" of a player is unknown to the game master, and may even be difficult to quantify, particularly in games with binary outcomes (wins/losses) rather than scores. We thus require a way to model skill, and a way to improve an existing model as the player plays more games. Early on in a player's career, we may wish to experiment with different opponents to learn more about the player's potential; however, if we are not able to quickly refine the quality of the matches, the player will leave the system. Approximate Bayesian inference provides a way to estimate player skill as well as model the uncertainty inherent in that estimate [8, 16].

*Bayesian logistic regression.* It is well known [24] that ordinary linear regression can be combined with Bayesian statistics to enable modelers to quantify the uncertainty of the regression estimators and make forecasts about the likely values of the true coefficients. This model also has the benefit of a fast, recursive updating scheme. There are many applications where we would like to have such a model, but for *logistic* regression: for example, consider the dynamic pricing application with a large number of explanatory variables representing features of the product and/or customer [31]. Unfortunately, there is no exact recursive update for logistic regression in either frequentist or Bayesian statistics. We discuss how approximate Bayesian inference can be used to create one.

## 3 Learning with conjugate Bayesian priors

In this section, we review several standard Bayesian models that enable sequential learning in a computationally efficient manner. We describe the property of conjugacy and explain why it is desirable, before moving on to models where this property does not hold. See DeGroot [10] or Powell and Ryzhov [29] for a more detailed exposition of this material.

Let $Y$ be a random observation drawn from a continuous distribution with density $g$, parameterized by a vector $\mu$. We write the density as $g(\cdot;\mu)$ to indicate this dependence. Suppose that $\mu$ itself is unknown. We adopt the Bayesian perspective and model $\mu$ as a random vector with joint density function $f(\cdot;\theta)$, parameterized by a vector $\theta$.

Note that these two probability distributions represent very different concepts. The *sampling density* $g$ represents exogenous uncertainty; the variance of $Y$ represents the amount of noise in the underlying real-world process from which the observation is generated (for example, the variation in one day of sales). We are trying to learn the distribution of $Y$ in order to understand this real-world process, which is equivalent to learning the values of $\mu$.

By contrast, the *prior density* $f$ is purely an object of belief, and the parameters $\theta$ are specified by the decision-maker. The mean of this distribution represents our point estimate of the unknown parameters $\mu$ of the sampling density; the variance-covariance matrix represents our uncertainty about the possible values that $\mu$ can take. The probability $P(\mu \in A)$ represents our belief that the unknown values are contained within some set $A$, whereas the probability

$$P(Y \in B) = \int P(Y \in B \,|\, \mu = u) f(u;\theta)\, du$$

is a forecast about the observation, based on our beliefs.

Suppose now that $Y$ is observed. Due to the noise of this observation, we will not be able to learn the exact value of $\mu$, but we can use the observation to update our distribution of belief. The *posterior* density of $\mu$ is simply the conditional density given $Y = y$, which can be calculated using Bayes' rule:

$$h(u \mid y) = \frac{g(y; u) f(u)}{\int g(y; v) f(v)\, dv}. \tag{1}$$

In some cases, it is possible to choose $f$ and $g$ in such a way that both the prior density $f(\cdot)$ and the posterior density $h(\cdot \mid y)$ belong to the same family of distributions. In this case, the posterior density $h(\cdot \mid y)$ can be rewritten as $f(\cdot; \theta')$, the original belief density evaluated using a set of updated posterior parameters $\theta'$, which depend on both $\theta$ (the old beliefs) and $y$ (the new information).

This property is known as *conjugacy*, and it considerably simplifies the learning process. Suppose now that we collect a sequence of observations $Y^1, Y^2, \ldots$. We can iteratively apply (1) to obtain a sequence of posterior densities $h^1, h^2, \ldots$. However, with conjugacy, this sequence takes the form $f(\cdot; \theta^1), f(\cdot; \theta^2), \ldots$, and we only need to store and update a finite-dimensional parameter vector $\theta^n$, representing the updated beliefs after $n$ observations. For standard conjugate models, one can derive a closed-form recursive update for $\theta^{n+1}$ as a function of $\theta^n$ and $Y^{n+1}$.

We now list several standard models where such updates are available. We focus specifically on models involving normal distributions, which provide an easy way to model correlations between elements of $Y$ or $\mu$ (when these are random vectors). However, we note that non-normal models are also subject to all of the issues discussed in this chapter; see Powell and Ryzhov [29] for some examples of non-normal conjugate models.

### *Univariate normal observation with known variance*

In the simplest possible example, suppose that $Y$ is normally distributed with mean $\mu$ (a scalar) and known, constant variance $\lambda^2$. The prior distribution of $\mu$ is also univariate normal with two parameters: the prior mean $\theta$, and the prior variance $\sigma^2$. Thus, the sampling density $g$ is parameterized by a single value $\mu$, while the prior density $f$ has two user-specified parameters.

Suppose that we observe $Y = y$. In this case, the numerator of (1) is given by

$$g(y; u) f(u; \theta, \sigma) = \frac{1}{\sqrt{2\pi\lambda^2}} e^{-\frac{(y-u)^2}{2\lambda^2}} \cdot \frac{1}{\sqrt{2\pi\sigma^2}} e^{-\frac{(u-\theta)^2}{2\sigma^2}}. \tag{2}$$

The first term on the right-hand side of (2) is the likelihood of observing $Y = y$, given that the unknown mean $\mu$ is exactly equal to $u$. The second term is then the likelihood that $\mu = u$, under the belief parameters $\theta, \sigma^2$. A simple exercise in algebra will show that (1) evaluates to

$$h(u \mid y) = \frac{1}{\sqrt{2\pi(\sigma')^2}} e^{-\frac{(u-\theta')^2}{2(\sigma')^2}}, \tag{3}$$

where

$$\theta' = \frac{\sigma^{-2}\theta + \lambda^{-2}y}{\sigma^{-2} + \lambda^{-2}}, \tag{4}$$

$$\left(\sigma'\right)^{-2} = \sigma^{-2} + \lambda^{-2}. \tag{5}$$

Since (3) is clearly a normal density, we can rewrite it as $f(u;\theta',\sigma')$, where the posterior parameters are easily computed from (4)–(5). Note that, in this case, $\theta'$ is a simple weighted average of the old point estimate $\theta$ and the new information $y$, with the weights given by the reciprocals of the prior and sampling variances. Thus, a low sampling variance $\lambda^2$ indicates that we should place more weight on the new information, whereas a low $\sigma^2$ indicates that we already had a high level of confidence in our old beliefs, and should stay closer to $\theta$.

We briefly note that this model is closely related to the standard method of estimating a population mean in classical statistics. Suppose that observations $Y^1, Y^2, \ldots$ are made sequentially, and (4)–(5) are iteratively applied to obtain $(\theta^1,\sigma^1)$, $(\theta^2,\sigma^2)$, and so on. Suppose that our initial prior variance, before any observations are collected, is given by $\left(\sigma^0\right)^2 = \infty$, implying that we start with no knowledge about the unknown mean. In this case, (4) will overwrite our prior point estimate $\theta^0$ after the first observation, and it can be shown that

$$\theta^n = \frac{1}{n}\sum_{i=1}^{n} Y^i, \qquad \sigma^n = \frac{\lambda^2}{\sqrt{n}},$$

reproducing the standard sample mean and its variance. This example provides the intuition that our point estimates of the unknown values behave quite similarly under Bayesian and frequentist models; the Bayesian model simply provides a way to make probabilistic forecasts about $\mu$.

*Univariate normal observation with unknown variance*

Now, consider a version of the same model where the sampling variance $\lambda^2$ is unknown. It is convenient to write $\rho = \lambda^{-2}$ and then model $\rho$ as a random variable. Thus, in this model, the sampling density $g$ is parameterized by both $\mu$ and $\rho$. It follows that $f$ is a bivariate density representing our uncertainty about these two quantities.

A standard conjugate model for this setting first assumes that $\rho$ follows a gamma distribution with parameters $a$ and $b$. Then, the conditional distribution of $\mu$, given that $\rho = r$, is assumed to be normal with mean $\theta$ and variance $\frac{1}{\tau r}$. The parameters $\theta$, $\tau$, $a$, and $b$ are all user-specified, and the prior density $f(u,r)$ can be written as the product of the marginal density of $\rho$ and the conditional density of $\mu$, given by

$$f(u,r;\theta,\tau,a,b) = \frac{1}{\sqrt{2\pi\tau^{-1}r^{-1}}}\frac{b\,(br)^{a-1}e^{-br}}{\Gamma(a)}e^{-\frac{(u-\theta)^2}{2\tau^{-1}r^{-1}}}, \tag{6}$$

where $\Gamma$ denotes the gamma function. The conditional sampling density, given $\mu = u$ and $\rho = r$, is simply

$$g(y; u, r) = \frac{1}{\sqrt{2\pi r^{-1}}} e^{-\frac{(y-u)^2}{2r^{-1}}}.$$

After more algebra, it can be shown that the posterior density $h(u \mid y)$ from (1) has the form $f(u; \theta', \tau', a', b')$, where

$$\begin{aligned} \theta' &= \frac{\tau\theta + y}{\tau + 1}, \\ \tau' &= \tau + 1, \\ a' &= a + \frac{1}{2}, \\ b' &= b + \frac{\tau (y - \theta)^2}{4(\tau + 1)}. \end{aligned}$$

In this way, a sequence of scalar observations can be used to simultaneously learn an unknown mean and variance.

As in the previous example, this model exhibits intuitive analogies to frequentist statistics. If we integrate (6) with respect to $r$, the marginal distribution of $\mu$ can be related to a Student's $t$-distribution. In fact, the quantity $\sqrt{\frac{\tau a}{b}}(\mu - \theta)$ follows a standard $t$-distribution with $2a$ degrees of freedom. As in frequentist statistics, we use $t$-distributions in place of normal distributions when the sampling variance is unknown.

*Multivariate normal prior, univariate normal observation with known variance*

We now consider an example with multiple dimensions. Let $Y$ be a random vector taking values in $\mathbb{R}^M$, and following a multivariate normal distribution with mean vector $\mu$. We assume that the covariance matrix of $Y$ is diagonal, and let $\lambda_i^2$ denote its $i$th diagonal entry. In this example we assume that all the variances $\lambda_i^2$ are known, and we suppose that $\mu$ is the only unknown parameter. Next, we suppose that $\mu$ itself follows a multivariate normal distribution with mean vector $\theta$ and covariance matrix $\Sigma$. The covariance matrix need not be diagonal. Thus, the prior density is given by

$$f(u; \theta, \Sigma) = \frac{1}{\sqrt{(2\pi)^M |\Sigma|}} e^{-\frac{1}{2}(u-\theta)^\top \Sigma^{-1}(u-\theta)},$$

where $u$ is now an $M$-vector.

In this model, we suppose that information is collected in the form of *univariate* samples. That is, instead of observing the full vector $Y$, we instead observe a scalar quantity $Y_i$ for some $i$. Clearly $Y_i \sim \mathcal{N}(\mu_i, \lambda_i^2)$. Applying (1) again, the posterior density $h(u \mid y)$ given $Y_i = y$ is multivariate normal with parameters

$$\theta' = \theta + \frac{y - \theta_i}{\lambda_i^2 + \Sigma_{ii}} \Sigma e_i, \tag{7}$$

$$\Sigma' = \Sigma - \frac{\Sigma e_i e_i^\top \Sigma}{\lambda_i^2 + \Sigma_{ii}}, \tag{8}$$

where $e_i$ is an $M$-vector of zeroes with only the $i$th component equal to 1. If $\Sigma$ is diagonal, (7)–(8) reduce to (4)–(5) applied to $\theta_i$ and $\Sigma_{ii}$. However, for general $\Sigma$, we now have the ability to change every component of $\theta$ on the basis of new information about just one of the component values.

In high dimensions, this model greatly increases the effect of a single piece of information. If the true values $\mu_i$ are believed to be strongly correlated, observing any one of those values may provide information about the entire vector. However, this places a greater burden on the user, who now has to specify an entire covariance matrix as part of the belief distribution. Note that the exogenous noise in this model is entirely uncorrelated, since we never sample two component values simultaneously. The matrix $\Sigma$ is an object of belief that models the "similarities" (or "differences") between the unknown components of $\mu$.

*Multivariate normal prior, multivariate normal observation*

The previous two examples can be extended to accommodate correlations in the sampling distribution. As before, we suppose that $Y$ is multivariate normal with mean $\mu$, but allow a general covariance matrix $\Lambda$. Suppose that both $\mu$ and $\Lambda$ are unknown. We can use a multivariate analog of the normal-gamma distribution described earlier. Suppose that the inverse $R = \Lambda^{-1}$ is a random matrix following a Wishart distribution [14]. The density of $R$ is given by

$$p\left(R; b, B\right) = \frac{1}{Z\left(b, B\right)} \left|R\right|^{\frac{b-M-1}{2}} e^{-\frac{1}{2}\mathrm{tr}(BR)},$$

where

$$Z\left(b, B\right) = \pi^{\frac{M(M-1)}{4}} \left|\frac{B}{2}\right|^{-\frac{b}{2}} \prod_{i=1}^{M} \Gamma\left(\frac{b+1-i}{2}\right) \tag{9}$$

is a normalizing constant. The Wishart density is a generalization of the gamma distribution to random matrices. It is convenient to place a belief distribution of $\Lambda^{-1}$, just as we did for the reciprocal of the sampling variance in the univariate case. Since this is a distribution of belief about the possible values of the sampling covariance matrix, the parameters $b, B$ are user-specified; $B$ is a matrix, while $b$ is a scalar.

The conditional sampling density, given $\mu = u$ and $\Lambda^{-1} = R$, is then assumed to be multivariate normal with mean vector $\theta$ and covariance matrix $\frac{1}{q}R^{-1}$, where $q, \theta$ are again user-specified ($\theta$ is a prior mean vector, while $q$ is a scalar). The joint density of $(\mu, R)$ is referred to as a normal-Wishart distribution. Thus, the decision-maker has to specify four distinct types of parameters, given by $q, b, \theta, B$. Given $Y = y$, the conditional distribution of $(\mu, R)$ remains normal-Wishart with updated parameters

$$q' = q + 1, \tag{10}$$

$$b' = b + 1, \tag{11}$$

$$\theta' = \frac{q\theta + Y}{q+1}, \tag{12}$$

$$B' = B + \frac{q}{q+1}(\theta - y)(\theta - y)^{\top}. \tag{13}$$

These equations provide intuition for the belief parameters. The two scalars $q$ and $b$ essentially function as sample sizes (they are incremented by 1 after every new observation). The vector $\theta$ represents point estimates of the unknown means, exactly as in the previous example. It is also analogous to a vector of sample means. Finally, $B$ represents a multivariate version of the empirical sum of squared errors. It can be shown that, under the normal-Wishart prior, $\mathbb{E}(\Lambda) = \frac{B}{b-M-1}$, quite similar to the usual sample covariance matrix.

*Multivariate normal prior in Bayesian linear regression*

Finally, we present a conjugate model for learning in linear regression; see Minka [24] for the full derivation. Here, we suppose that the observation $Y$ is a scalar of the form

$$Y = \beta^{\top} x + \varepsilon, \tag{14}$$

where $x \in \mathbb{R}^M$ is a fixed vector of regression features (known to the decision-maker), and $\varepsilon \sim \mathcal{N}(0, \lambda^2)$ is an independent noise with known variance. The regression coefficients $\beta$ are unknown. The Bayesian model assumes that $\beta$ follows a multivariate normal distribution with mean vector $\theta$ and covariance matrix $C$; these are the only belief parameters in the model. Then, the conditional distribution of $\beta$, given $Y = y$ and the associated vector $x$, is still multivariate normal with parameters

$$\theta' = \theta + \frac{y - \theta^{\top} x}{\lambda^2 + x^{\top} C x} C x, \tag{15}$$

$$C' = C - \frac{C x x^{\top} C}{\lambda^2 + x^{\top} C x}. \tag{16}$$

These updating equations generalize (7)–(8), with $e_i$ replaced by the feature vector $x$. In this setting, even if the prior covariance matrix $C$ is diagonal, the posterior may incorporate covariances since multiple features are being combined into a single scalar observation.

As in the preceding examples, this model essentially replicates the behavior of frequentist least squares. It is easy to see that (15)–(16) are essentially identical to the well-known recursive least squares update [28, Sect. 9.3]. Thus, $\theta$ is essentially the usual least squares estimator of the regression coefficients. However, as before, the Bayesian model adds the dimension of uncertainty quantification, allowing us to make probabilistic forecasts about likely values of $\beta$.

## 4 Conjugacy in simulation and optimization

In traditional Bayesian statistics, conjugacy may not be necessary. If we suppose that information is collected once, in a batch rather than sequentially, it is only necessary to solve a single estimation problem and identify the posterior distribution of the unknown parameters given the entire set of samples. In such cases, Markov chain Monte Carlo methods, which are computationally intensive but enjoy theoretical convergence guarantees, may be used.

Conjugacy becomes much more valuable when information is collected sequentially, and especially when the decision-maker can influence the information collection process. Consider the following simple example. Let $\mu_i$ be unknown values, for $i = 1, \dots, M$. Each $\mu_i$ is modeled as a random variable, but we assume that $\mu_i$ and $\mu_j$ are independent for any $i \neq j$. At every stage of sampling, we have the ability to observe $Y_i \sim \mathcal{N}\left(\mu_i, \lambda_i^2\right)$ for any $i$. However, we can only choose to observe one $i \in \{1, \dots, M\}$ at a time.

In this problem, information comes in the form of a sequence $Y^1_{i^0}, Y^2_{i^1}, \dots$, where $i^0, i^1, \dots$ is the sequence of sampling choices. The unknowns $\mu_i$ in this problem may represent the performance values of $M$ different *alternatives*. For example, $\mu_i$ may be the mean power output of a wind farm built in the $i$th candidate location, the mean throughput of the $i$th possible factory layout, or the mean sales observed under the $i$th pricing strategy. To collect a single observation, we may either run a time-consuming simulation (e.g., a discrete-event model of a factory layout), or perform a field experiment (e.g., implement a pricing strategy on our website and see what happens). In many situations, it is not feasible to experiment with every alternative simultaneously, giving rise to the new optimization problem of choosing the sequence $i^1, i^2, \dots$ in order to efficiently learn about some quantity of interest. For example, if we wish to maximize the mean power output of our wind farm, we would be interested in finding $\arg\max_i \mu_i$, the highest-valued alternative.

Due to our independence assumptions, learning occurs in this problem by iterative applying (4)–(5) to the belief parameters for whichever alternative has just been sampled. Let us initialize our beliefs by assuming that $\mu_i \sim \mathcal{N}\left(\theta_i^0, \left(\sigma_i^0\right)^2\right)$. Then, we can write

$$\theta_i^{n+1} = \begin{cases} \frac{\left(\sigma_i^n\right)^{-2}\theta_i^n + \lambda^{-2} Y_{i^n}^{n+1}}{\left(\sigma_i^n\right)^{-2} + \lambda_i^{-2}} & \text{if } i^n = i \\ \theta_i^n & \text{if } i^n \neq i, \end{cases} \tag{17}$$

$$\left(\sigma_i^{n+1}\right)^{-2} = \begin{cases} \sigma^{-2} + \lambda_i^{-2} & \text{if } i^n = i \\ \left(\sigma_i^n\right)^{-2} & \text{if } i^n \neq i. \end{cases} \tag{18}$$

Suppose now that our budget for information collection is limited to $N$ observations, where $N$ is very small (and can even be smaller than $M$, the number of alternatives). To maximize the effectiveness of each sample, we would like the ability to choose $i^n$ adaptively, based on the most current information. Thanks to conjugacy, our beliefs after $n$ samples are completely characterized by two quantities $\theta_i^n$ and $\sigma_i^n$ for each $i$. To design an adaptive sampling allocation, we can simply choose some function of these parameters. For instance,

$$i^n = \arg\max_i \theta_i^n + z \cdot \sigma_i^n \tag{19}$$

allocates the next sample based on an "optimistic" estimate. The point estimate $\theta_i^n$ is augmented by an "uncertainty bonus" that grows with $\sigma_i^n$, the idea being that higher $\sigma_i^n$ makes it more likely that the true value $\mu_i$ is substantially better than the point estimate. The quantity $z$ may be a tunable parameter representing the weight given to uncertainty relative to the point estimate.

Note that (19) defines a simple, yet adaptive algorithm. After each new sample, we will update our beliefs using (17)–(18), which may change the alternative chosen by (19) for the next sample. Of course, the performance of this algorithm (known under the name of interval estimation; see Kaelbling [19]) will heavily depend on the value of the tunable parameter $z$. The design of such algorithms is an active area of research, and these issues are outside the scope of this chapter; see Hong and Nelson [17] or Chau et al. [4] for an introductory overview from the perspective of the simulation community.

For our purposes, we can observe that conjugacy greatly simplifies the design of such adaptive algorithms. The sampling decision $i^n$ is based on $M$ continuous distributions of belief about the alternatives; however, due to conjugacy, these distributions can be characterized by a small set of parameters. Consequently, algorithms such as (19) can be simple, closed-form computations with finitely many inputs. Both learning and optimization in this problem can thus be performed very quickly. It would be much more difficult to proceed without a way to concisely characterize the belief distributions at every stage; nonetheless, this exact issue arises in numerous applications, motivating the use of approximate Bayesian inference.

## 5 Approximate Bayesian inference

Approximate Bayesian inference becomes applicable when the posterior density $h(u \mid y)$ in (1) no longer belongs to the same family as the prior density $f$. This occurs when there is a mismatch between the prior and the sampling density $g$.

For example, suppose that $Y \sim \mathcal{N}\left(\mu, \lambda^2\right)$ and that we have a normal prior on $\mu$, but instead of observing $Y$ directly, we observe a binary signal $1_{\{Y \geq y\}}$ for some $y$. Now, in the right-hand side of (1), the numerator is a product of a normal density and a normal cdf, which cannot be made to resemble a normal density through algebraic manipulations.

This creates substantial complications for sequential learning, particularly when the belief density at every stage of sampling is an input to an optimization algorithm, as in Section 4. Suppose that, in the above example, the level $y$ in the observation $1_{\{Y \geq y\}}$ is actually chosen by the decision-maker. If we attempt to iteratively apply (1), the posterior density will become more and more complicated after each stage. Even for a simple algorithm, such as the one in (19), it may become quite difficult to compute simple means and variances for such a density. It is also not clear how the density should be stored, if we cannot concisely describe it with a few parameters.

In such a situation, we may wish to assume that the posterior belongs to the same family as the prior. More specifically, we may choose to replace the "difficult" posterior density $h(\cdot \mid y)$ by one of the form $f\left(\cdot; \tilde{\theta}\right)$, where we choose $\tilde{\theta}$ to minimize some measure of the difference between these two densities (or, to put it another way, to maximize the similarity between them). We then simply discard $h$ and proceed with the next stage of sampling based on $f\left(\cdot; \tilde{\theta}\right)$ only. Applying (1) to the next sample will again lead to a difficult posterior, which can again be approximated in the same way. Thus, we make a tradeoff between the accuracy of our statistical model and the computational convenience of designing learning algorithms, which can once more be defined in terms of functions of the approximate parameters $\tilde{\theta}$, rather than the entire belief density.

Although approximate Bayesian models are difficult to analyze theoretically, they are very powerful computationally, and allow one to solve difficult sequential problems that would otherwise be difficult to approach using Bayesian models. Below, we review two useful computational techniques for approximate Bayesian inference, moment matching and density filtering, and discuss how these techniques have been implemented in the recent literature to solve problems of interest.

## 5.1 Moment matching

In the first method, we choose values of $\tilde{\theta}$ that solve the system

$$\int u^k f\left(u; \tilde{\theta}\right) du = \int u^k h(u \mid y)\, du, \tag{20}$$

for $k = 1, \ldots, K$ with some desired $K$. Essentially, this method sets the approximate parameters $\tilde{\theta}$ in such a way that the approximate belief density $f\left(\cdot; \tilde{\theta}\right)$ has the same moments (up to the $K$th order) as the actual posterior $h(\cdot \mid y)$, where $y$ is the given value of the next observation. We may expect larger values of $K$ to lead to more accuracy, but in practice we take $K$ to be large enough that the system in (20) has a unique solution. For example, if the desired density $f$ has two parameters, we may take $K = 2$ and obtain those parameters by matching the mean and variance.

The equations in (20) are nonlinear and may be difficult to solve. However, in some applications, a closed-form, easily computable solution is available. We now discuss some examples from the recent literature.

*Application to competitive online gaming*

Recall the competitive gaming application discussed in Section 2, where an online gaming system attempts to match opponents in a competitive event based on their perceived skill level. Herbrich et al. [16] and Dangauthier et al. [8] present the following Bayesian model for learning unknown player skills. Let $\mu_i$ denote an abstract "skill value" for player $i$. This value may not be directly reflected in the competition, and does not have any units; skill is determined based on the relative magnitudes of these values, so that player $i$ is more skillful than $j$ on average if $\mu_i > \mu_j$. Define a random variable $Y_i \sim \mathcal{N}\left(\mu_i, \lambda^2\right)$, where $\lambda^2$ is assumed known, to represent the "performance" of player $i$ in any given game. In the actual event, performance is measured in terms of wins and losses, rather than a numerical score, so $Y_i$ is again an abstract representation of performance allowing for some random variation around the player's average skill level.

Consider a game where players compete one-on-one. We do not directly observe their performance values. However, if player $i$ wins, we interpret this as a signal that $Y_i > Y_j$. Similarly, if $j$ wins, this means that $Y_j < Y_i$ (we assume that draws do not happen in this game). Since we do not know either player's skill level, we begin with the assumption that $\mu_i \sim \mathcal{N}\left(\theta_i, \sigma_i^2\right)$, representing a prior belief. We also assume that all skill levels $\mu_i$ are mutually independent. However, the information in this problem takes the form $1_{\{Y_i > Y_j\}}$. This random variable is non-normal and so (1) will not yield a normal posterior.

In this application, it is especially crucial to be able to model and update the beliefs efficiently. In online gaming, there are typically thousands of players waiting to be matched at any given moment, and any match-making algorithm must quickly map the belief parameters to a matching decision. Fortunately, (20) can be solved in closed form for $K = 2$ moments, leading to the update

$$\tilde{\theta}_i = \begin{cases} \theta_i + \frac{\sigma_i^2}{\sigma_i^2+\sigma_j^2+2\lambda^2} \cdot v\left(\frac{\theta_i - \theta_j}{\sigma_i^2+\sigma_j^2+2\lambda^2}\right) & \text{if } Y_i > Y_j, \\ \theta_i - \frac{\sigma_i^2}{\sigma_i^2+\sigma_j^2+2\lambda^2} \cdot v\left(\frac{\theta_j - \theta_i}{\sigma_i^2+\sigma_j^2+2\lambda^2}\right) & \text{if } Y_i < Y_j, \end{cases} \tag{21}$$

and

$$\tilde{\sigma}_i^2 = \begin{cases} \sigma_i^2\left(1 - \frac{\sigma_i^2}{\sigma_i^2+\sigma_j^2+2\lambda^2} \cdot w\left(\frac{\theta_i - \theta_j}{\sigma_i^2+\sigma_j^2+2\lambda^2}\right)\right) & \text{if } Y_i > Y_j, \\ \sigma_i^2\left(1 - \frac{\sigma_i^2}{\sigma_i^2+\sigma_j^2+2\lambda^2} \cdot w\left(\frac{\theta_j - \theta_i}{\sigma_i^2+\sigma_j^2+2\lambda^2}\right)\right) & \text{if } Y_i < Y_j, \end{cases} \tag{22}$$

where

$$v(x) = \frac{\phi(x)}{\Phi(x)},$$
$$w(x) = v(x)(v(x) + x),$$

and $\phi$, $\Phi$ denote the standard normal pdf and cdf. The moment-matching problem only approximates the marginal posterior distribution of $\mu_i$: note that the observation in this problem depends on both $Y_i$ and $Y_j$, which should potentially induce correlations in the posterior distribution. However, for computational convenience, these correlations are simply dropped from the model (intuitively, they may not be significant due to the large pool of players), and we continue to assume that all skill levels are independent and normally distributed, with belief parameters given by (21)–(22).

In this way, our beliefs about a player are completely characterized by two numbers which can be updated via a quick recursive calculation after each new game played. One attractive feature of this model is that the approximate update (21)–(22) behaves fairly intuitively. If $i$ beats $j$, we increase our point estimate of $\mu_i$; if not, we decrease the estimate. The belief variance goes down over time as more games are played, implying that our beliefs become more accurate when we have observed player $i$ for a period of time.

A simple algorithm for matching players (and one that is used in practice, as discussed in Herbrich et al. [16]) might calculate

$$P\left(\left|Y_i - Y_j\right| < \delta\right) \approx 2\delta \frac{1}{\sqrt{2\pi\left(\sigma_i^2 + \sigma_j^2 + 2\lambda^2\right)}} e^{-\frac{\left(\theta_i - \theta_j\right)^2}{2\left(\sigma_i^2 + \sigma_j^2 + 2\lambda^2\right)}}, \tag{23}$$

which represents the likelihood that the game will end in a "draw" (i.e., that the players are evenly matched), for every pair of available players. Then, the game master may choose the match that maximizes this criterion in a myopic fashion. While this algorithm may not be theoretically optimal, it is very easy to compute based on the parameters of the normal belief distribution. Of course, if player $i$ has already played some games, we will simply plug the parameters of the approximate Bayesian model into (23). We now have a fast way to make decisions, thanks to the computational simplifications afforded by the approximate model. In fact, Herbrich et al. [16] describe a field implementation of such an algorithm at a major online video gaming service.

*Application to market design*

In this application, we consider a market where every trader interacts with a market-maker who participates in every transaction, but may be either the buyer or the seller.

Our exposition here is based on the results of Das and Magdon-Ismail [9], where it is assumed that each transaction involves a fixed purchase quantity. See Brahma et al. [2] for an extension to variable volumes.

Suppose that a single asset is traded (one unit at a time) in the market. Let $\mu$ denote the market value of this asset. This value is unknown and is subject to high uncertainty (for example, if the market has just experienced a shock). The market-maker keeps a Bayesian prior $\mu \sim \mathcal{N}\left(\theta, \sigma^2\right)$.

Traders arrive and interact with the market-maker one at a time. Let $Y$ denote a trader's perception of the unknown value. Given $\mu$, the conditional distribution of $Y$ is normal with mean $\mu$ and variance $\lambda^2$. A trader is thus analogous to a single "observation" in our previous examples. The variance $\lambda^2$ represents variation among the population of traders. For simplicity, let us assume that this variance is known. Let $s = \frac{\sigma}{\lambda}$ be the ratio of the two variances in the problem; this quantity is useful for modeling uncertainty in this application.

The market-maker functions as both buyer and seller, by specifying a "bid price" $b$ and an "ask price" $a$. The trader can either buy one unit of the asset at $a$ or sell at $b$. However, the trader will only buy if $Y < b$ (the trader thinks that the asset is over-valued), and likewise will only sell if $Y > b$. If $a \leq Y \leq b$, the trader does nothing. The market-maker thus observes a censored signal $\bar{y} \in \{1, 0, -1\}$ which indicates that the trader bought, idled, or sold, respectively. It is clear that (1) does not yield a normal posterior, for the same reasons as in the online gaming application.

However, moment-matching can be applied. Define some notation

$$z^+ = \begin{cases} \infty & \text{if } \bar{y} = 1 \\ a & \text{if } \bar{y} = 0 \\ b & \text{if } \bar{y} = -1 \end{cases}, \qquad z^- = \begin{cases} a & \text{if } \bar{y} = 1 \\ b & \text{if } \bar{y} = 0 \\ -\infty & \text{if } \bar{y} = -1 \end{cases}.$$

It can then be shown that moment-matching yields the posterior parameters

$$\tilde{\theta} = \theta + \sigma \frac{B}{A}, \tag{24}$$

$$\tilde{\sigma}^2 = \sigma^2 \left(1 - \frac{AC + B^2}{A^2}\right), \tag{25}$$

where the quantities $A, B, C$ are given by

$$A = I\left(\frac{z^+ - \theta}{\lambda}, s\right) - I\left(\frac{z^- - \theta}{\lambda}, s\right),$$

$$B = J\left(\frac{z^+ - \theta}{\lambda}, s\right) - J\left(\frac{z^- - \theta}{\lambda}, s\right),$$

$$C = K\left(\frac{z^+ - \theta}{\lambda}, s\right) - K\left(\frac{z^- - \theta}{\lambda}, s\right),$$

with

$$I(\alpha,\beta) = \Phi\left(\frac{\alpha}{\sqrt{1+\beta^2}}\right),$$

$$J(\alpha,\beta) = -\sqrt{\frac{\beta^2}{1+\beta^2}}\phi\left(\frac{\alpha}{\sqrt{1+\beta^2}}\right),$$

$$K(\alpha,\beta) = \frac{\alpha\beta^2}{(1+\beta^2)^{\frac{3}{2}}}\phi\left(\frac{\alpha}{\sqrt{1+\beta^2}}\right).$$

The full derivation is given in Das and Magdon-Ismail [9] and we do not repeat it here. We note, however, that (24)–(25) provide a way to process thousands of interactions in a very short time. It can furthermore be shown that $\tilde{\sigma}^2 \leq \sigma^2$, implying that uncertainty goes down over time (as in the conjugate case, or in the online gaming model).

In fact, the properties of the approximate Bayesian model have implications beyond computational convenience. With this model in hand, one may wish to specify an objective for the market-maker. Das and Magdon-Ismail [9] consider two objectives, one where the market-maker seeks to maximize cumulative discounted profit, and one where the market-maker seeks to make zero profit (break even while increasing the liquidity of the market). In both cases, the market-maker's decision consists of setting bid and asking prices before each interaction. However, due to the symmetry of the normal distribution, it can be shown that the optimal choices for these prices are symmetric around the current posterior mean. Thus, if $(\theta,\sigma)$ are our current belief parameters, we let $a = \theta + \delta$ and $b = \theta - \delta$ for some $\delta$. An optimal policy for setting $\delta$ can be described using Bellman's equation [28],

$$V(\sigma) = \max_{\delta} R(\sigma,\delta) + \gamma \mathbb{E}\left[V(\tilde{\sigma}) \mid \delta\right], \tag{26}$$

where $R$ is the one-period expected profit function (expected revenue from a trader who buys minus expected cost from a trader who sells), and $0 < \gamma < 1$ is a discount factor. Since our beliefs are completely characterized by a mean and a variance, these parameters should also be sufficient to make a decision at any given time. Note, however, that the value function in (26) only depends on $\sigma$, not on the mean (it is not at all obvious that this should be the case, but nonetheless it can be shown). Since the (approximate) belief variance decreases over time, it is easier to design a search procedure that will solve (26) efficiently.

Of course, one may expect the approximation to become less accurate as we run more iterations, since additional error is incurred every time the approximate update is performed. However, the empirical evidence in Das and Magdon-Ismail [9] suggests that a normal distribution approximates the true posterior reasonably well.

The true posterior appears to be unimodal, and normal approximations often work well in such cases [13]. While it is difficult to theoretically prove the consistency of the posterior mean (see Section 6 for a more detailed discussion of the challenges involved), the empirical evidence also suggests that the Bayesian market-maker does indeed learn the correct value of $\mu$ over time. See also Chakraborty et al. [3] for a case application of the market-maker in a prediction market with human agents.

*Application to dynamic pricing*

Our last application of moment-matching is motivated by the problem of learning demand curves in dynamic pricing of digital goods, studied by Chhabra and Das [5]. A seller has infinite supply of a homogeneous good with zero production cost (a reasonable simplification for digital goods such as audio files) and offers a price $p$ to a prospective buyer, who either accepts (thus earning a revenue of $p$ for the seller) or rejects (resulting in zero revenue). We assume that the customer has a certain valuation $Y$ of the good, and accepts if and only if $Y \geq p$. In the absence of detailed data about customers, it is common to assume that $Y$ is uniformly distributed on an interval $[0, B]$, where $B$ is unknown. This suggests that the customers are homogeneous, but keeps some uncertainty about them.

As a consequence, the conditional probability (given $B$) that the customer accepts is given by $1 - \beta p$, where $\beta = \frac{1}{B}$. Since this is a linear function of $p$, we refer to this model as "linear demand." For simplicity, we normalize the valuations and assume that $\beta \in [0, 1]$. In this case, a normal belief distribution is not appropriate, but we may assume that $\beta$ follows a beta distribution with parameters $a$ and $b$. That is,

$$f(u; a, b) = \frac{\Gamma(a+b)}{\Gamma(a)\,\Gamma(b)} u^{a-1} (1-u)^{b-1}.$$

Let $F(\cdot; a, b)$ be the beta cdf for fixed $a, b$. The information that we observe takes the form $\tilde{y} = 1_{\{Y \geq p\}}$ for some $p$ that we may assume fixed in this discussion. Working out (1) will show that $h(u \mid y)$ cannot be represented as a beta density.

However, moment-matching may be used to create an approximate posterior $f\left(\cdot; \tilde{a}, \tilde{b}\right)$. Chhabra and Das [5] show that the approximate parameters solve a certain system of nonlinear equations. It is, however, necessary to derive separate updates for the case where the price is accepted and the case where it is not accepted. If the price is accepted and the customer buys, one can show that

$$\frac{\tilde{a}}{\tilde{a}+\tilde{b}} = \frac{p\mathbb{E}(\beta^2)\, F\left(\frac{1}{p}; a+2, b\right) + \mathbb{E}(\beta)\left(1 - F\left(\frac{1}{p}; a+1, b\right)\right)}{p\mathbb{E}(\beta)\, F\left(\frac{1}{p}; a+1, b\right) + 1 - F\left(\frac{1}{p}; a, b\right)},$$

$$\frac{\tilde{a}(\tilde{a}+1)}{\left(\tilde{a}+\tilde{b}\right)\left(\tilde{a}+\tilde{b}+1\right)} = \frac{p\mathbb{E}(\beta^3)\, F\left(\frac{1}{p}; a+3, b\right) + \mathbb{E}(\beta^2)\left(1 - F\left(\frac{1}{p}; a+2, b\right)\right)}{p\mathbb{E}(\beta)\, F\left(\frac{1}{p}; a+1, b\right) + 1 - F\left(\frac{1}{p}; a, b\right)}.$$

On the other hand, if the customer does not buy, one can derive the system

$$\frac{\tilde{a}}{\tilde{a}+\tilde{b}} = \frac{\mathbb{E}(\beta)\,F\left(\frac{1}{p};a+1,b\right) - p\mathbb{E}\left(\beta^2\right)F\left(\frac{1}{p};a+2,b\right)}{F\left(\frac{1}{p};a,b\right) - p\mathbb{E}(\beta)\,F\left(\frac{1}{p};a+1,b\right)},$$

$$\frac{\tilde{a}\,(\tilde{a}+1)}{\left(\tilde{a}+\tilde{b}\right)\left(\tilde{a}+\tilde{b}+1\right)} = \frac{\mathbb{E}\left(\beta^2\right)F\left(\frac{1}{p};a+2,b\right) - p\mathbb{E}\left(\beta^3\right)F\left(\frac{1}{p};a+3,b\right)}{F\left(\frac{1}{p};a,b\right) - p\mathbb{E}(\beta)\,F\left(\frac{1}{p};a+1,b\right)}.$$

These equations can be inverted to obtain

$$\tilde{a} = \frac{M_1 M_2 - M_1^2}{M_1^2 - M_2},$$

$$\tilde{b} = \frac{(1-M_1)\,\tilde{a}}{M_1},$$

where $M_1, M_2$ are the first two moments of the exact posterior $h(\cdot \mid \tilde{y})$. This example illustrates the use of moment-matching with non-normal beliefs. Chhabra and Das [5] show that the resulting model is competitive against several standard benchmarks from the dynamic pricing literature.

## *5.2 Density filtering*

The method of moment matching may work well when the exact and approximate posteriors have a similar shape (e.g., if both are unimodal). In some cases, however, we may wish to approximate the overall shape of the posterior distribution as well as possible, rather than simply match its mean. The technique of density projection or density filtering [25] may be used for this purpose. For fixed $\tilde{\theta}$, define

$$\mathscr{D}^{KL}(f\,||\,h) = \mathbb{E}_f\left(\log\frac{f\left(\mu;\tilde{\theta}\right)}{h(\mu \mid y)}\right) \tag{27}$$

to be the Kullback-Leibler divergence between the exact posterior $h(\cdot \mid y)$ and the approximate density $f\left(\cdot;\tilde{\theta}\right)$. Observe that, in (27), $\mu$ is a random variable, assumed to have the density $f\left(\cdot \mid \tilde{\theta}\right)$ in *both* the numerator and denominator (indicated by the notation $\mathbb{E}_f$). We are thus taking the expected logarithm of the Radon-Nikodym derivative $\frac{df}{dh}$, viewed as a function of $\mu$. It can be shown that this quantity is bounded below by zero, and equals zero only if $f$ and $h$ are identical. Thus, the KL divergence may be viewed as a measure of the "distance" between two probability distributions: if it is greater, the distributions will be less similar.

If (27) can be evaluated for fixed $\tilde{\theta}$, we can then optimize it by solving the problem

$$\min_{\tilde{\theta}} \mathscr{D}^{KL}(f \,||\, h), \tag{28}$$

thus finding the set of posterior parameters that make the approximate density $f\left(\cdot; \tilde{\theta}\right)$ maximally similar to the actual posterior. If (28) can be solved efficiently, we proceed exactly as with moment matching: at each stage of sampling, we calculate the posterior parameters and then discard the actual posterior, assuming instead that $f\left(\cdot; \tilde{\theta}\right)$ is the correct belief density and working with that density for further computation and optimization.

The order of $f$ and $h$ in (27) may be reversed, with $h$ in the numerator (and with the expectation taken over $h$). This representation can also be viewed as a measure of distance, but is not equivalent to the one in (27) and may produce a different set of parameters. The choice between $\mathscr{D}^{KL}(f \,||\, h)$ and $\mathscr{D}^{KL}(h \,||\, f)$ may be made based on computational convenience. Since $h$ is typically difficult to handle, while $f$ is some common distribution chosen for its tractability, (28) is likely to be easier to solve. We now discuss two examples where efficient solutions are available.

*Application to learning unknown correlation structures*

Recall the normal-Wishart learning model from Section 3. The observation $Y$ is an $M$-vector following a multivariate normal distribution with mean vector $\mu$ and covariance matrix $\Lambda$, both unknown. We assume that $R = \Lambda^{-1}$ follows a Wishart distribution with scalar parameter $b$ and matrix parameter $B$. Then, the conditional distribution of $Y$ given $R$ is multivariate normal with mean vector $\theta$ and covariance matrix $\frac{1}{q}R^{-1}$. The joint distribution of $(\mu, R)$ has four belief parameters $(q, b, \theta, B)$.

As discussed above, when $Y$ is observed, the posterior distribution is normal-Wishart. Now, however, suppose that we only observe a single component $Y_i$ of $Y$, for some arbitrary $i$. We are thus attempting to combine the normal-Wishart model with the second example from Section 3, where a scalar observation was used to update a multivariate belief distribution. In that case, however, we assumed a known covariance matrix, whereas the normal-Wishart model considers an unknown $R$. It turns out that the conditional distribution of $(\mu, R)$ given $Y_i$ is no longer normal-Wishart. Rather, it is a mixture of a normal-Wishart and normal distribution that cannot be expressed in terms of any one common family of distributions. If we are collecting a sequence of scalar observations, as in Section 4, approximate Bayesian inference may help us model the learning process concisely.

**Theorem 1 ([30]).** *In the normal-Wishart model, the KL divergence defined in (27), given $Y_i = y$, can be written in closed form as*

$$\begin{aligned}\mathscr{D}^n_{KL}(f||h) = & \frac{\tilde{b}-b}{2}\left(-\log\left|\frac{\tilde{B}}{2}\right| + \sum_{j=1}^{M}\psi\left(\frac{\tilde{b}-j+1}{2}\right)\right) - \frac{\tilde{b}M}{2} \\ & + \frac{\tilde{b}}{2} tr\left(B(\tilde{B})^{-1}\right) + \log\frac{Z(b,B)}{Z(\tilde{b},\tilde{B})} + \frac{1}{2}\log\tilde{B}_{ii} \\ & + \frac{1}{2}\left[M\log\frac{\tilde{q}}{q} + M\frac{q}{\tilde{q}} - M + q(\theta-\tilde{\theta})^\top\tilde{b}(\tilde{B})^{-1}(\theta-\tilde{\theta})\right] \\ & - \frac{1}{2}\psi\left(\frac{\tilde{b}-M+1}{2}\right) + \frac{1}{2\tilde{q}} + \frac{1}{2}(y-\tilde{\theta}_i)^2\frac{\tilde{b}-M+1}{\tilde{B}_{ii}} + \kappa,\end{aligned}$$

*where $\psi(x) = d\log\Gamma(x)/dx$ is the digamma function, $Z$ is as in (9), and $\kappa$ is a constant that does not depend on the parameters of $f$.*

Since the KL divergence can be written in closed form, it is easier to optimize it. The KL divergence is convex in the belief parameters, so it is sufficient to take first derivatives with respect to the belief parameters and set them equal to zero. As it turns out, the resulting equations are linked only through the scalar parameter $\tilde{b}$, leading to the following result.

**Theorem 2 ([30]).** *There exists a finite value $\Delta b$ such that the optimal solution to (28) can be expressed as*

$$\tilde{q} = q + \frac{1}{M}, \tag{29}$$

$$\tilde{b} = b + \Delta b, \tag{30}$$

$$\tilde{\theta} = \theta + \frac{y-\theta_i}{\frac{q\tilde{b}}{\tilde{b}-M+1}B_{ii} + B_{ii}}Be_i, \tag{31}$$

$$\tilde{B} = \frac{\tilde{b}}{b}B + \frac{\tilde{b}}{b+1}\left(\frac{q\,(y-\theta_i)^2}{\frac{q\tilde{b}}{\tilde{b}-M+1}+1} - \frac{B_{ii}}{b}\right)\frac{Be_ie_i^\top B}{B_{ii}^2}. \tag{32}$$

This is not exactly a closed-form solution, but is still fairly straightforward to solve. Equations (29)–(32) provide closed-form updates for fixed $\Delta b$. Then, $\Delta b$ itself can be found by applying a standard optimization method (e.g., gradient descent) to the first derivative of $\mathscr{D}^{KL}(f\,||\,h)$ with respect to $\tilde{b}$. Since this is a one-dimensional optimization problem, it can be solved much more quickly than attempting to apply gradient methods to the KL divergence directly.

The approximate update exhibits interesting parallels to (7)–(8). As in the conjugate model, the change in the belief vector $\theta$ is driven by the scalar difference $y - \theta_i$.

The observation may influence every component of $\theta$ through the correlations modeled in the matrix $B$. Since the noise of the observations is unknown, $B$ is simultaneously used to model possible correlations in the exogenous process represented by $Y$, as well as correlations in the distribution of belief representing perceived similarities between components of $Y$.

Recall that, in the conjugate model, the matrix $B$ is analogous to an empirical sum of squares. However, if only scalar observations can be collected, the empirical correlations are not accessible, and we use the scalar deviation $(y - \theta_i)^2$, combined with the correlations already present in $B$, to stand in for these quantities. Finally, the scalar parameters are approximated in a way that supports their interpretation as a sample size. In the conjugate model, $q$ is incremented by 1 after each vector observation; however, in this model, we only observe one out of $M$ components, so the increment becomes $\frac{1}{M}$. The increment $\Delta b$ does not have a closed form, but typically falls between 0 and $\frac{1}{M}$, and was observed empirically [32] to converge to $\frac{1}{M}$.

The approximate update exhibits intuitively "correct" behavior, and leads to good empirical performance [30], but proving its correctness (e.g., in the sense of statistical consistency) is an open problem. Section 6 discusses the theoretical challenges arising in approximate Bayesian inference.

*Application to Bayesian logistic regression*

Recall from Section 3 that ordinary least-squares regression can be viewed from a Bayesian perspective. When the observation has the form $Y = \beta^\top x + \varepsilon$ for fixed $x$ and random, zero-mean $\varepsilon$, we may place a multivariate normal prior on $\beta$ and update it using (15)–(16). This elegant model can then be used in conjunction with sequential simulation and optimization methods [15, 26].

It would be very convenient to have a similar model for *logistic* regression. Consider a dynamic pricing problem in which products and customers are heterogeneous. As in Section 5.1, a seller first chooses a price $p$, and then observes the buyer's binary response $Y$, which is equal to 1 if the product was purchased and 0 otherwise. Now, however, we use the logistic demand model [36]

$$P(Y = 1) = \frac{1}{1 + e^{-\beta^\top x}}, \tag{33}$$

where $x$ is a vector of regression features (including the price). In the simplest possible model, we have $x = [1, p]^\top$, meaning that customers and products are homogeneous. However, in practice, $x$ may be an $M$-vector that includes other features that model characteristics of products and customers. For example, we may include dummy variables that check for customers in a certain region, or specific types of products, or interaction terms between the two.

We would like to use a multivariate normal prior $\beta \sim \mathcal{N}(\theta, C)$, as in the linear model. As we have seen, the normal distribution provides a very simple way to model correlations, and correlations are crucial in regression, since they are automatically induced when our observation is a blend of multiple coefficients.

We then rewrite (33) as

$$P(Y=1) = \ell(H(\beta)),$$

where $\ell(z) = \frac{1}{1+e^{-z}}$ and $H(\beta) = (2Y-1)\left(\beta^\top x\right)$. Then, (1) becomes

$$h(u \mid y) \propto \ell(H(\beta))\,|C|^{-\frac{1}{2}}\, e^{-\frac{1}{2}(\beta-\theta)^\top \Sigma^{-1}(\beta-\theta)},$$

which is evidently not normal. However, the following computational result suggests that density filtering may again be useful.

**Theorem 3 ([31]).** *Given Y, the KL divergence can be written as*

$$\mathscr{D}^{KL}(f \,||\, h) = \mathbb{E}_f\left[\log\left(1+e^{-H(\beta)}\right)\right] + h\left(\theta, C, \tilde{\theta}, \tilde{C}\right), \tag{34}$$

*where*

$$h\left(\theta, C, \tilde{\theta}, \tilde{C}\right) = \frac{1}{2}\left[tr\left(C^{-1}\tilde{C}\right) + \left(\theta-\tilde{\theta}\right)^\top C^{-1}\left(\theta-\tilde{\theta}\right) - M - \log\frac{|\tilde{C}|}{|C|} + \kappa\right]$$

*and $\kappa$ is a constant that does not depend on the parameters of the approximate posterior.*

The KL divergence can be partially simplified; however, (34) includes an expectation that does not have a closed form. We now have two options. First, since (28) optimizes an expectation, we may apply gradient-based stochastic optimization [21] to optimize the expectation in (34). This approach is considered by Blei et al. [1] using a likelihood-ratio estimator [34] of the gradient of the difficult expectation. However, without further simplification, such methods may take a long time to find a good solution due to the high dimensionality of our decision variables (both $\tilde{\theta}$ and $\tilde{C}$). For this reason, we consider some other techniques from the literature.

Since Bayesian linear regression admits a conjugate model and is generally well-understood, many authors have sought to connect the conjugate update of (15)–(16) to the logistic setting. Consider the model

$$\tilde{\theta} = \theta + \frac{\left(Y-\frac{1}{2}\right)v - \theta^\top x}{v + x^\top C x} Cx, \tag{35}$$

$$\tilde{C} = C - \frac{Cxx^\top C}{v + x^\top C x}. \tag{36}$$

In Bayesian linear regression, the numerator in (35) would simply be $y - \theta^\top x$, with $y$ being the continuous response. In logistic regression, the observation is binary, so we first subtract $\frac{1}{2}$ (to make positive values indicate successes and negative values indicate failures) and scale the result by a factor $v$. The presence of $v$ in the denominator of (35) suggests that it is meant to stand in for the residual noise in the linear regression model. Since the logistic model does not have such a parameter, $v$ is artificial and should be somehow chosen by the decision-maker.

This approximation does not optimize (28), but it has two benefits. First, it is quite easy to understand and makes an analogy between the logistic model and the much more well-structured linear model. Second, if we adopt (35)–(36) as our update, the problem of choosing $\left(\tilde{\theta}, \tilde{C}\right)$ for our approximate posterior reduces to the problem of choosing a single scalar parameter $v$. For this reason, previous authors have used this "linearization" of the problem due to its computational convenience. For instance, Spiegelhalter and Lauritzen [35] experiment with

$$v = \hat{p}\,(1 - \hat{p}),$$

where $\hat{p}$ is simply the sample mean of the observations collected thus far. This approach relies on the analogy between $v$ and a sample variance. Another approximation by Jaakkola and Jordan [18] proposes

$$v^{-1} = \frac{\frac{1}{2} - \ell(\xi)}{2\xi}$$

and recursively updates

$$\xi = x^\top \tilde{C} x + \left(x^\top \tilde{\theta}\right)^2$$

after every iteration. This approach is based on a Taylor approximation to the non-normal posterior, providing a rigorous foundation for the use of an update that resembles linear regression.

A third approach, adopted by Qu et al. [31], is to simply choose $v$ to optimize (28) subject to the additional constraint that $\left(\tilde{\theta}, \tilde{C}\right)$ satisfy (35)–(36). As a consequence, (34) still contains a difficult integral, but $v$ is now the only parameter to optimize, which makes it much easier to use gradient-based methods. In experiments, this approach appears to be competitive with the closed-form update of Jaakkola and Jordan [18]. In all of these cases, we obtain an efficient recursive update for Bayesian logistic regression that carries over much of our intuition about linear regression.

## 6 Theoretical challenges

So far, it seems clear that approximate Bayesian inference offers significant computational benefits for sequential learning problems, particularly when information collection is itself optimized using algorithms that need to run fast. The theoretical properties of approximate Bayesian models are much less straightforward. Unfortunately, even standard statistical properties such as asymptotic consistency do not easily lend themselves to tractable analysis under posterior approximation.

We briefly describe the challenges using consistency as an example. All of the conjugate models in Section 3 are consistent, meaning that the estimated means converge to the true values asymptotically. For instance, consider the model with normal observations and known variance, where we collect observations $Y^1, Y^2, \ldots$ and construct a sequence $(\theta^n, \sigma^n)$ of posterior parameters. In this model, $\theta^n \to \mu$ almost surely, which becomes obvious when we consider the analogy between $\theta^n$ and a frequentist sample mean. In the simulation literature, Bayesian algorithms rely on the consistency of the estimators as a given; the question is whether the algorithms collect enough information about every alternative for the asymptotic convergence to take place.

In studying consistency in Bayesian models, it is standard to invoke martingale arguments. First, in the aforementioned conjugate model, we have $\theta^n = \mathbb{E}(\mu \mid \mathscr{F}^n)$ by definition, where $\mathscr{F}^n$ is the sigma-algebra generated by $Y^1, \ldots, Y^n$. That is, $\theta^n$ is always the conditional mean of $\mu$ given $\mathscr{F}^n$. Since every observation is an unbiased estimate of $\mu$, it follows that $\mu$ is always centered around $\theta^n$, even though the value of $\theta^n$ may change between iterations. It follows [7] that $(\theta^n)$ is uniformly integrable and converges almost surely to $\mu$.

This does not happen in approximate Bayesian inference. Consider the approximate normal-Wishart model in (29)–(32). By analogy to the conjugate case, we can let $\left(\tilde{\theta}^n\right)$ be the sequence of posterior means obtained by applying the approximate updating equations after observing $Y^1_{i^0}, \ldots, Y^n_{i^{n-1}}$. But it is no longer the case that $\mathbb{E}\left(\mu \mid Y^1_{i^0}, \ldots, Y^n_{i^{n-1}}\right)$ equals $\theta^n_i$. In fact, we have no way of knowing what $\mathbb{E}\left(\mu \mid Y^1_{i^0}, \ldots, Y^n_{i^{n-1}}\right)$ is; to do so, we would have to calculate the actual posterior distribution after $n$ observations, which may not be analytically tractable. We may still assume that $Y_i$ is an unbiased observation of $\mu_i$, but we no longer have uniform integrability, or a way to identify the limit of $\left(\tilde{\theta}^n\right)$.

Our intuition may then be to apply martingale arguments to a model where the approximate Bayesian assumptions hold. That is, supposing that $\mu$ really does follow a normal-Wishart distribution given $Y^1_{i^0}, \ldots, Y^n_{i^{n-1}}$, we may perhaps be able to reconstruct the standard arguments. But this essentially replaces the original probability measure in the problem (let us denote it as $\mathbb{P}$) by a different measure $\tilde{\mathbb{P}}$, under which $\mu$ may be centered around $\tilde{\theta}^n$, but it is no longer possible to guarantee that $Y$ is centered around $\mu$. Thus, in approximate Bayesian inference, we can either assume that $\mu$ is centered around $\tilde{\theta}$, or that $Y$ is centered around $\mu$, but not both. Unfortunately, it is quite difficult to connect the approximate measure $\tilde{\mathbb{P}}$ back to the original measure $\mathbb{P}$ after multiple observations have been made. For instance, if the two probability measures could be shown to be equivalent, we could then argue that they have the same almost sure events (so consistency under one measure also holds in the other). Unfortunately, since each measure is induced by a distribution for the infinite sequence $\left(\mu, R, Y^1_{i^0}, Y^2_{i^1}, \ldots\right)$, it is difficult to demonstrate equivalence. In general, equivalence for two infinite-dimensional probability measures can only be shown in a few special cases [11], none of which apply to the normal-Wishart setting.

Thus, martingale arguments are largely inapplicable. We may then consider a second widely used theoretical technique. Here, one would view the approximate update (29)–(32) as a certain dynamical system (perhaps with unbiasedness assumptions on the stochastic observation), and invoke ODE theory or stochastic approximation results to obtain convergence of the iterate $\tilde{\theta}^n$. See Kushner and Yin [22] for an exposition of this approach, and Chien and Fu [6] for an application that showed consistency of the conjugate normal-Wishart model. Unfortunately, the standard ODE results tend to assume that the dynamical system is linear, which clearly does not hold for (29)–(32), and it is difficult to apply it otherwise without very restrictive assumptions on the observations (such as boundedness, which excludes normal distributions).

Perhaps for these reasons, consistency of approximate Bayesian models is also largely an open problem in the statistical literature. Recent work [20, 27] has studied the asymptotic behavior of the approximate posterior parameters, but such results typically require very strong assumptions. For instance, Kim [20] shows that the KL divergence (essentially the distance between $\mathbb{P}$ and $\tilde{\mathbb{P}}$) vanishes to zero at a certain rate, but first requires consistency of the posterior parameters (also with a certain convergence rate) as an assumption. Thus, even limited consistency results for some of the models considered in this chapter would represent substantial theoretical progress for the field. In the meantime, the computational advantages of approximate Bayesian inference continue to drive practical, efficient learning models and algorithms.

## 7 Conclusion

We have described two techniques of approximate Bayesian inference and illustrated their use in several recent applications where information is collected sequentially and used in optimal learning algorithms. We have mainly emphasized the practical benefits of this methodology – in all of the surveyed applications, approximate Bayesian models considerably simplified the representation and refinement of decision-makers' beliefs. These models provide attractive solutions in broad problem classes where Bayesian inference is otherwise difficult to apply, such as logistic regression.

This chapter has not considered every possible approximation technique. For instance, another popular technique is the maximum a posteriori or MAP method, where the decision-maker attempts to design a sequence of posterior densities that converge asymptotically to a point mass on the true value; see, e.g., Garcia-Fernandez and Svensson [12] for recent work on this approach. Nonetheless, we hope that the variety of applications surveyed here serves as evidence in favor of the flexibility and practical import of the approximate Bayesian methodology.

## References

1. Blei, D.M., Jordan, M.I., Paisley, J.W.: Variational Bayesian inference with stochastic search. In: Proceedings of the 29th International Conference on Machine Learning, pp. 1367–1374 (2012)
2. Brahma, A., Chakraborty, M., Das, S., Lavoie, A., Magdon-Ismail, M.: A Bayesian market maker. In: Proceedings of the 13th ACM Conference on Electronic Commerce, pp. 215–232 (2012)
3. Chakraborty, M., Das, S., Lavoie, A., Magdon-Ismail, M., Naamad, Y.: Instructor rating markets. In: Proceedings of the 27th AAAI Conference on Artificial Intelligence, pp. 159–165 (2013)
4. Chau, M., Fu, M.C., Qu, H., Ryzhov, I.O.: Simulation optimization: a tutorial overview and recent developments in gradient-based methods. In: Tolk, A., Diallo, S.Y., Ryzhov, I.O., Yilmaz, L., Buckley, S., Miller, J.A. (eds.) Proceedings of the 2014 Winter Simulation Conference, pp. 21–35 (2014)
5. Chhabra, M., Das, S.: Learning the demand curve in posted-price digital goods auctions. In: Proceedings of the 10th International Conference on Autonomous Agents and Multi-Agent Systems, pp. 63–70 (2011)
6. Chien, Y.T., Fu, K.-S.: On Bayesian learning and stochastic approximation. IEEE Trans. Syst. Sci. Cybern. **3**(1), 28–38 (1967)
7. Cinlar, E.: Probability and Stochastics. Springer, New York (2011)
8. Dangauthier, P., Herbrich, R., Minka, T.P., Graepel, T.: TrueSkill through time: revisiting the history of chess. In: Platt, J.C., Koller, D., Singer, Y., Roweis, S. (eds.) Advances in Neural Information Processing Systems, Curran Associates, Inc., vol. 20, pp. 337–344 (2007)
9. Das, S., Magdon-Ismail, M.: Adapting to a market shock: optimal sequential market-making. In: Koller, D., Bengio, Y., Schuurmans, D., Bottou, L., Culotta, R. (eds.) Advances in Neural Information Processing Systems, Curran Associates, Inc., vol. 21, pp. 361–368 (2008)
10. DeGroot, M.H.: Optimal Statistical Decisions. Wiley, New York (1970)
11. Engelbert, H.J., Shiryaev, A.N.: On absolute continuity and singularity of probability measures. Banach Cent. Publ. **6**, 121–132 (1980)
12. Garcia-Fernandez, A.F., Svensson, L.: Gaussian MAP filtering using Kalman optimisation. IEEE Trans. Autom. Control, **60**(5):1336–1349 (2015)
13. Gelman, A., Carlin, J., Stern, H., Rubin, D.: Bayesian Data Analysis, 2nd edn. CRC Press, Boca Raton (2004)
14. Gupta, A., Nagar, D.: Matrix Variate Distributions. Chapman & Hall/CRC, London (2000)
15. Han, B., Ryzhov, I.O., Defourny, B.: Efficient learning of donor retention strategies for the American Red Cross. In: Pasupathy, R., Kim, S.-H., Tolk, A., Hill, R., Kuhl, M.E. (eds.) Proceedings of the 2013 Winter Simulation Conference, pp. 17–28 (2013)
16. Herbrich, R., Minka, T.P., Graepel, T.: TrueSkill™: a Bayesian skill rating system. In: Schölkopf, B., Platt, J.C., Hoffman, T. (eds.) Advances in Neural Information Processing Systems, MIT Press, vol. 19, pp. 569–576 (2006)
17. Hong, L.J., Nelson, B.L.: A brief introduction to optimization via simulation. In: Rosetti, M., Hill, R., Johansson, B., Dunkin, A., Ingalls, R. (eds.) Proceedings of the 2009 Winter Simulation Conference, pp. 75–85 (2009)

18. Jaakkola, T.S., Jordan, M.I.: Bayesian parameter estimation via variational methods. Stat. Comput. **10**(1), 25–37 (2000)
19. Kaelbling, L.P.: Learning in Embedded Systems. MIT Press, Cambridge (1993)
20. Kim, J.-Y.: Limited information likelihood and Bayesian analysis. J. Econom. **107**(1), 175–193 (2002)
21. Kim, S.: Gradient-based simulation optimization. In: Perrone, L.F., Wieland, F.P., Liu, J., Lawson, B.G., Nicol, D.M., Fujimoto, R.M. (eds.) Proceedings of the 2006 Winter Simulation Conference, pp. 159–167 (2006)
22. Kushner, H.J., Yin, G.: Stochastic Approximation and Recursive Algorithms and Applications, 2nd edn. Springer, Berlin (2003)
23. Marmidis, G., Lazarou, S., Pyrgioti, E.: Optimal placement of wind turbines in a wind park using Monte Carlo simulation. Renew. Energy **33**(7), 1455–1460 (2008)
24. Minka, T.P.: Bayesian linear regression. Technical report, Microsoft Research (2000)
25. Minka, T.P.: A family of algorithms for approximate Bayesian inference. Ph.D. thesis, Massachusetts Institute of Technology (2001)
26. Negoescu, D.M., Frazier, P.I., Powell, W.B.: The knowledge-gradient algorithm for sequencing experiments in drug discovery. INFORMS J. Comput. **23**(3), 346–363 (2010)
27. Pati, D., Bhattacharya, A., Pillai, N.S., Dunson, D.: Posterior contraction in sparse Bayesian factor models for massive covariance matrices. Ann. Stat. **42**(3), 1102–1130 (2014)
28. Powell, W.B.: Approximate Dynamic Programming: Solving the Curses of Dimensionality, 2nd edn. Wiley, New York (2011)
29. Powell, W.B., Ryzhov, I.O.: Optimal Learning. Wiley, New York (2012)
30. Qu, H., Ryzhov, I.O., Fu, M.C.: Ranking and selection with unknown correlation structures. In: Laroque, C., Himmelspach, J., Pasupathy, R., Rose, O., Uhrmacher, A.M. (eds.) Proceedings of the 2012 Winter Simulation Conference, pp. 144–155 (2012)
31. Qu, H., Ryzhov, I.O., Fu, M.C.: Learning logistic demand curves in business-to-business pricing. In: Pasupathy, R., Kim, S.-H., Tolk, A., Hill, R., Kuhl, M.E. (eds.) Proceedings of the 2013 Winter Simulation Conference, pp. 29–40 (2013)
32. Qu, H., Ryzhov, I.O., Fu, M.C., Ding, Z.: Sequential selection with unknown correlation structures. Operations Research, **63**(4):931–948 (2015)
33. Ryzhov, I.O., Tariq, A., Powell, W.B.: May the best man win: simulation optimization for match-making in e-sports. In: Jain, S., Creasey, R.R., Himmelspach, J., White, K.P., Fu, M.C. (eds.) Proceedings of the 2011 Winter Simulation Conference, pp. 4239–4250 (2011)
34. Spall, J.C.: Introduction to Stochastic Search and Optimization: Estimation, Simulation, and Control. Wiley-Interscience, Hoboken (2005)
35. Spiegelhalter, D.J., Lauritzen, S.L.: Sequential updating of conditional probabilities on directed graphical structures. Networks **20**(5), 579–605 (1990)
36. Talluri, K.T., Van Ryzin, G.J.: The Theory and Practice of Revenue Management. Springer, New York (2006)

# Optimization for Power Systems and the Smart Grid

**Miguel F. Anjos**

**Abstract** Practitioners in the field of power systems operations are keen users of optimization techniques. Several fundamental problems in the area are solved every day using optimization algorithms as part of the real-time operation of the power grid. One such fundamental problem is the unit commitment problem that is concerned with scheduling power generation so as to meet demand at minimum cost. Realistic instances of unit commitment are typically large-scale, and because the time available in the real-time context is limited, practitioners sometimes have to settle for solutions that are not globally optimal. Beyond the well-known fundamental problems, the advent of the smart grid introduces new challenges for power system researchers and for optimizers. The smart grid is the combination of a traditional electrical power distribution system with two-way communication between suppliers and consumers. This combination is expected to deliver energy savings, cost reductions, and increased reliability and security. However it also raises new difficulties for managing of the resulting system. These include integrating renewable energy sources such as wind and solar power generation, managing bidirectional flows of power and of information, and incorporating demand-response. This chapter begins with an overview of the area of smart grid and some of the challenges relevant to optimization researchers. We then summarize two recent examples of optimization research in power systems, the first consisting of an application to demand-response for the smart grid, and the second of a new technique to solve certain types of unit commitment more efficiently.

**Keywords** Unit commitment • Smart grid • Optimization • Demand reponse • Power systems

**MSC (2010):** 90C11, 90C15, 90B30

---

M.F. Anjos (✉)
Canada Research Chair and Director, Trottier Institute for Energy, Polytechnique Montreal & GERAD, Montreal, QC, Canada
e-mail: anjos@stanfordalumni.org



B. Defourny, T. Terlaky (eds.), *Modeling and Optimization: Theory and Applications*, Springer Proceedings in Mathematics & Statistics 147,
DOI 10.1007/978-3-319-23699-5_2

## 1 Introduction

This chapter is based on the plenary presentation with the same title that was delivered at MOPTA 2014. The purpose of both the presentation and this chapter is to increase awareness within the community of optimization researchers of the breadth of opportunities to contribute in the area of power systems operations.

Power systems practitioners are keen users of optimization techniques, and several fundamental problems in the area are solved daily using optimization algorithms. The changing needs of practitioners motivate the continuing research to improve the optimization tools to solve these problems. One such fundamental problem is the unit commitment problem that has been studied since at least the 1960s. Unit commitment is concerned with scheduling power generation so as to meet demand at minimum cost. Realistic instances of unit commitment are typically large-scale and require significant computational time to solve, and because the time available in the context of system operations is limited, practitioners sometimes have to settle for solutions that are not globally optimal.

Beyond these fundamental problems, the advent of the smart grid introduces new challenges for power system researchers and for optimizers. The term "smart grid" has become ubiquitous in the power systems area. There are now annual conferences about it, including the *IEEE PES Conference on Innovative Smart Grid Technologies* and *Smart Grid Canada*. The IEEE Power & Energy Society sponsors a Transactions journal on the subject, and a number of reports and technical standards have been released or are in development.

We begin by giving a high-level description of what "smart grid" is, from the author's point of view. The word "grid" clearly refers to the electricity network that transports electricity from where it is generated to where it is consumed. This means carrying electricity from (typically large-scale) generating units to industrial, commercial, and residential consumers.

The word "smart" is less precise. In this context it generally refers to the changes that are happening, or need to take place, for the power system to better meet the needs and expectations of society in the 21st century. In this paper we focus on the context of large grids in developed countries, as the issues are significantly different in other contexts, such as isolated sites, islands, or developing countries.

The term *smart grid* is thus a convenient way to encapsulate a number of unfolding developments concerning the electricity system and that are part of a broader push for energy efficiency. These developments include:

- Power distribution companies always had to send people out to gather the data needed to manage their operations, such as consumption, voltage levels, and equipment status. It is now possible to deploy data-gathering devices such as smart meters, voltage sensors, and fault detectors, together with the two-way communication technology to transmit the data automatically as well as to control the network components remotely. Similar changes have mostly already taken place at the power transmission level via the use of technologies such as Supervisory Control And Data Acquisition (known as SCADA) [16] and Phasor

Measurement Units (known as PMUs) [21]. While the extent to which the power system has been automated and computerized varies from one jurisdiction to another, there is an irreversible trend in the electricity industry to do so.

- Economic growth is accompanied by a corresponding growth in the demand for electricity but generation has often failed to keep up. The result is a reduction (and sometimes near absence) of spare supply, and hence a tightly constrained operating context. The moments of high power consumption levels are called *load peaks* and they are a major concern for the system operator. In some instances, the growth in peak demand has in fact surpassed the growth in annual demand for electricity, and the new capacity has struggled to keep up. While such extreme situations currently occur only during a few days of the peak season (e.g., winter in Quebec or summer in New York), we may be heading towards having overall tighter operating margins and higher capacity factors of the installed capacity, meaning that generation will be closer to the system's maximum possible output.
- Concerns over environmental damage from the extraction and consumption of fossil fuels are leading to the integration of ever-increasing amounts of electricity from renewable sources into the power system. The intermittent nature of most of these sources, notably wind power and solar power, leads to important technical challenges to ensure the stability and reliability of the grid.
- Progress in energy storage technologies and power electronics have made it technically feasible to store energy in increasingly large quantities. As energy storage becomes commercially viable, grid-scale energy storage becomes an additional tool for power system planners, and its integration will have a tremendous impact on the operation of the grid. This is because energy storage systems can fulfill a number of important functions, including smoothing the output fluctuations from intermittent renewable generation, and supporting the grid operation during load peak periods by releasing energy stored at other times.
- Electric vehicles, although still negligible in number, will become a significant load on the grid as their numbers increase. They will become a major component of the load in the smart grid of the future, although their batteries are effectively energy storage devices and can thus in principle (though the practice is complicated!) fulfill some of the function of storage mentioned above.

Some of the consequent challenges are:

- The need to handle very large amounts of data and to unlock the value of these data. This is a context-specific application of what is currently termed *Big Data*.
- The need for demand to frequently adjust to the supply available (unlike traditionally when mostly the supply adjusted to the demand) so as to assist in achieving the necessary constant balance between supply and demand. This is referred to as *demand side management* or *demand response*.
- The need to move from a centralized, fully controllable grid to one that integrates large proportions of *distributed generation*, i.e., decentralized, intermittent generation, and that can manage large numbers of electric vehicles.

It turns out that optimization techniques are useful tools to address these and a number of other challenges arising from the above developments.

In the remainder of this chapter we illustrate the application of optimization in power systems operations by summarizing two recent optimization research projects by the author's research group. First, Section 2 gives an introduction to the architecture proposed by Costanzo et al. [7] for managing a system of diverse loads such as in a building, for example. The power consumption of buildings is important because worldwide it accounts for an estimated 40% of global energy consumption [27]. The potential for buildings to act as providers of demand-response is thus significant. The novelty of the architecture is its layered structure with communication interfaces for handling bidirectional information exchange with the loads and with the grid. The emphasis is on controlling the operation of loads to achieve the desired outcome in terms of energy consumption.

Second, Section 3 summarizes a novel technique to handle symmetry in mixed-integer linear optimization recently proposed by Ostrowski et al. [24]. This novel technique, called *Modified Orbital Branching*, is particularly relevant in the context of multiple power generators that have identical or near-identical operational features.

## 2 Autonomous Building Load Management

The first application we present is in the area of demand-response, also known as demand-side management. The concept of demand-side management was introduced in the 1970s in response to the rise of energy cost and the need to conserve energy. Early demand-side management was carried out via such techniques as direct load control [12] in which utilities can disconnect selected appliances when needed. Although direct load control is an effective approach for large loads, it is not practical for a large number of small loads. The latter is however the typical scenario in the case of buildings, where coordinating the operation of many, often very small, loads is necessary to manage the total energy consumption for the building.

One alternative proposed is the concept of a *smart building* where the appliances can be individually managed and the coordination of energy consumption is carried out locally. With this approach, the load control is handled locally by the consumer, with the utility influencing the consumer's decisions on power consumption via the energy price that is determined according to the energy market, the network load, and other economic and technical factors. One of the issues to address is that a system of loads of different magnitudes (such as a building) is characterized by loads with different magnitudes, heterogeneous dynamics, and multiple time-scales. Indeed appliance control is carried out in real-time, on a scale of at most a few minutes, whereas handling price bidding and scheduling of loads are performed in a longer time-scale, typically hours. We provide here an introduction to the system architecture proposed in [7] to manage the loads from appliances in so-called smart buildings. The purpose of this architecture is to manage the operation of appliances to achieve the desired outcome in terms of energy/power consumption.

## 2.1 System Architecture

To efficiently manage a set of heterogeneous loads, an appropriate classification of power consumption modes is needed. Following previous work in the literature, see, e.g., [4, 11, 17], we assume here that the loads are characterized according to the following three categories:

1. **Baseline load** is the consumption of appliances that must be served immediately at any time to keep them operating or on standby. This includes lighting, cooking stove, computing and network devices. Baseline load cannot be deferred and thus must be taken into account when computing the (remaining) available capacity to satisfy other demands at every point in time.
2. **Burst loads** correspond to appliances that run each time for a fixed duration and are required to start and finish within specified time periods. Examples of such appliances include washing machines, dryers, and dishwashers. Simultaneous operation of several burst loads contributes to the increase of peak loads, therefore careful management of burst loads has a significant impact on the stability of the power system.
3. **Regular loads** correspond to appliances that are always operational, such as heating, air conditioning, refrigerators, and water heaters. Unlike baseline load however, these appliances can be interrupted under certain conditions, and hence their operation can be managed. Note that from an operational perspective, regular loads can be viewed as a special case of burst loads.

The architecture proposed in [7] consists of three layers: the Admission Controller (AC), the Load Balancer (LB), and a third layer containing the Demand Response Manager (DRM) and the Load Forecaster (LF). The AC in the bottom layer performs real-time load control by interacting with the appliances via their individual interfaces. It uses a strategy inspired by scheduling in real-time computing systems, see [5] and the references therein. Operating requests that can be accepted right away are allowed to proceed, otherwise the request is rejected (or deferred) and passed on to LB.

The LB schedules the deferred load requests by solving a mixed-integer linear optimization (MILO) problem that minimizes the overall operational cost subject to the global capacity constraints for the building, and the operational requirements of each appliance. A typical objective for the LB will be to spread the load over the scheduling horizon so as to contribute to the reduction of the system peaks. Once the scheduling is done, the LB communicates to each appliance the time for it to again make a request to the AC. This next request may be accepted or it may be again deferred, depending on the circumstances at that time, and in particular depending on whether rescheduling has occurred in the meantime. Each reschedule is triggered by events such as the arrival of new requests, or changes in the power capacity limit for the building (or system in general).

This leads us to the need to eliminate the possibility that a request be deferred indefinitely. The implementation in [7] includes with each request a priority value between 0 and 1 representing the imperativeness of the task. For example, appliances such as a refrigerator or a water heater will have a value of 0 when their inner temperature is well within the pre-defined *comfort zone*, with the value increasing as the boundary of the comfort zone is approached, and reaching 1 when the temperature is at the boundary of the zone. For burst loads with deadlines, this priority value is also a function of the time remaining before the latest start time, defined as the latest moment at which the task can be started and still be able to complete on time.

At the top layer, the DRM handles the interaction between the system and the grid. It receives from the LB basic performance parameters, including capacity utilization and rejection rate. It also receives from the grid one or more capacity limits and the corresponding energy unit costs, and passes that information to the LB. The DRM tracks quality of service issues such as the rate of rejection of requests. Based on all this information, the DRM can react to price signals from the grid, or make requests to the grid for additional power capacity when necessary. This design allows the DRM to accommodate different pricing strategies, such as critical-peak, time-of-use, or real-time pricing.

The Load Forecaster (LF) is an auxiliary module in the top layer that provides the DRM and the LB with information to assist them in their tasks. For example, a reliable load forecast makes it possible to plan the operation of appliances in advance so as to avoid peak load periods on the grid, or take advantage of lower consumption periods when the energy price drops.

This system architecture has the following important properties:

- **Scalability**: It can be used in a variety of circumstances, ranging from private homes to commercial buildings, factories, campuses, military bases, and micro-grids. While the complexity of the components can vary significantly, the system structure remains the same.
- **Extensibility**: It supports the integration of intermittent renewable energy resources as well as of energy storage devices. The consequent changes in the optimization objectives and constraints are straightforward to make.
- **Composability**: The hierarchical nature of the architecture allows price bidding to be carried out at different levels, not only at the top level. In this way, different pricing strategies can be integrated, and even coexist, in the same system.

We summarize in the next section the MILO formulation of the LB used in [7]. For more details and a complete presentation of the system architecture, we refer the reader to that paper and to the experimental results in [6].

## *2.2 Load Balancer Formulation*

The LB spreads the load requests over a given time horizon in order to minimize a cost function related to the energy price while satisfying a limit on the power consumption at each time period, and the deadlines of the requests.

Each task is scheduled over a proper number of consecutive time frames in a finite scheduling horizon. Two basic assumptions for load balancing are:

- Each appliance draws a given amount of power when it operates;
- Energy cost and power capacity limit have been provided by the DRM.

To present the model formulation of load balancing, we consider a problem consisting of $n$ requests to be scheduled in a horizon containing $m$ equal time frames. We denote by $\mathcal{N} = \{1, \ldots, n\}$ and $\mathcal{M} = \{1, \ldots, m\}$ two index sets corresponding to the set of appliances and the time frames, respectively. For $i \in \mathcal{N}, j \in \mathcal{M}$, let $x_{ij}$ be a binary variable representing the activation state of the $i$th appliance in the $j$th time frame with 0 and 1 representing the states "inactive" and "active," respectively. If $P_i$ is the power consumption of appliance $i$ and $K_j$ is the energy cost per time unit, then $F_{ij} = P_i K_j$ defines a cost for appliance $i$ to operate during time frame $j$. Furthermore, for appliances operating over more than one time frame, we introduce additional variables, $d_{ij}, i \in \mathcal{N}, j \in \mathcal{M}$, that take the value 1 if appliance $i$ is scheduled to start at the time frame $j$. These variables force the allocation of consecutive time frames to this appliance once it starts via the constraints (3). In general, we can also associate with each $d_{ij}$ a startup cost denoted by $G_{ij}$. We can now state the MILO problem for the LB:

$$\min \sum_{i,j} F_{ij} x_{ij} + \sum_{i,j} G_{ij} d_{ij}, \tag{1}$$

$$\text{s.t.} \sum_{i} P_i x_{ij} \leq C_j, \quad \forall j \in \mathcal{M}, \tag{2}$$

$$d_{ij} \leq x_{it}$$

$$t = j, j+1, \cdots j + \tau_i - 1, \quad \forall j \in \mathcal{M}, \forall i \in \mathcal{N}, \tag{3}$$

$$\sum_{j} d_{ij} = 1, \quad \forall i \in \mathcal{N}, \tag{4}$$

$$x_{ij} = 0, \quad \forall i \in \mathcal{N}, \quad \forall j \notin (T_i^{earliest}, T_i^{latest}), \tag{5}$$

$$d_{ij} \geq 0, \quad \forall i \in \mathcal{N}, \quad \forall j \in \mathcal{M}, \tag{6}$$

$$x_{ij} \in \{0, 1\}, \quad \forall i \in \mathcal{N}, \quad \forall j \in \mathcal{M}, \tag{7}$$

where $C_j$ is the available capacity for time frame $j$, $\tau_i$ is defined as

$$\tau_i = \left\lceil \frac{E_i}{P_i} \right\rceil$$

which is the number of consecutive time frames for appliance $i$ that requires a total amount of energy $E_i$ for operation, and $T_i^{earliest}$ and $T_i^{latest}$ are, respectively, the earliest and latest start time of appliance $i$.

In this problem there are four sets of constraints, specifically:

1. The total power consumption at each time frame has to respect the given capacity limit (2).
2. For each request a proper number of contiguous time frames is allocated so that each appliance is operated for a sufficient period in order to complete the working cycle before the deadline (3).
3. Each task is scheduled only once (4). This number can be modified according to the task characteristics and requirements.
4. Each task is scheduled in an allowed operation period in such a way that each appliance is operated in a specific time interval (5).

An interesting feature of the LB formulation is that it captures the total power consumption of all the managed appliances. Consequently, it allows a simpler model of demand response than some others in the literature [4, 11, 26]. We observe also that this formulation is suitable for applications characterized by time-coupling and variable power requests, such as electric vehicle battery charging. Such a complex task can be split into a sequence of requests characterized by constant power demand with proper timing constraints, provided the appliance possesses the required level of intelligence. The advantage of such an approach is that it facilitates managing appliances with unpredictable behavior.

## 3 Symmetry in Unit Commitment

The second application we present is to one of the fundamental problems in the operation of power systems, namely the Unit Commitment (UC) problem. We say that a power generator is *committed* if it is scheduled by the system operator to provide power in given amounts over a given time period. The objective of UC is to find a power generation schedule for each generator in the system so as to meet demand at minimum cost, and to do so while ensuring that the grid operates safely and reliably. The fundamental UC problem may be stated as follows: Given a set of power generators and a set of electricity demands, the objective is to minimize the total production cost of the power generators subject to the constraints that

1. the demand is met, and
2. the generators operate within their physical limits, i.e.,
   a. the power output level of a generator may not change too rapidly (ramping constraints), and
   b. when a generator is turned on (off), it must stay on (off) for a minimum amount of time (minimum up / downtime constraints).

The survey paper [1] in the Proceedings of MOPTA 2012 provides an introduction to UC from the point of view of optimization and a summary of the recent developments in the literature.

Most system operators use MILO solvers on a daily basis to inform commitment decisions. In the real-world context of power system operation, the instances of UC that must be solved are often challenging because they are large-scale and require significant computational time to solve, while the time available to solve a UC model is a hard limitation.

The solution of UC problems can be particularly difficult when there are many identical generators, resulting in an optimization problem with many symmetries. A MILO problem exhibits symmetry if some or all of its variables can be permuted without changing the structure of the problem. Because symmetry increases the size of the search tree in algorithms to solve MILO problems, its presence typically leads to degraded computational performance. For this reason, there has been much research devoted to removing it using so-called symmetry-breaking techniques.

In the application to UC, symmetry shows up in the MILO formulation when several generating units have the same (or sufficiently similar) characteristics, and thus they can be permuted without changing the optimization problem. The integration of distributed generation not only increases the number of generating units in the system but also often introduces symmetry because several of them are typically permutable, particularly when they are based on the same technology. The current developments in smart grid thus increase the importance of methods to handle symmetry, because the performance of MILO solvers normally improves when effective symmetry-handling methods are used.

The past decade has seen major advances in general methods for symmetry breaking in MILO, including isomorphism pruning [19, 20] and orbital branching [22]. Furthermore several important classes of MILO problems contain highly structured symmetry groups that can be exploited. This observation has motivated the development of problem-specific techniques. For example, orbitopal fixing [14, 15] is an efficient technique to break symmetry in bin packing problems.

For the UC problem, one way to remove symmetry from the UC problem's formulation is to aggregate all identical generators into a single generator. However aggregating generator variables may be very difficult, and some of the physical requirements may be difficult to enforce [18]. Alternatively, we may consider using a general symmetry breaking-method that can take advantage of the special structure in UC. The latter was the approach illustrated in [24] where a novel technique called *Modified Orbital Branching (ModOB)* was shown to be particularly effective for the UC problem. ModOB is especially useful for problems whose solutions can be expressed as orbitopes. In this chapter we focus on the impact of ModOB for solving UC problems, and we refer the reader to [24] for a detailed study and theoretical analysis of ModOB.

## 3.1 Mathematical Formulation of the UC Problem with Symmetry

We present below a formulation of UC originally presented in [24] that explicitly models the presence of multiple generators with the same characteristics. The generators are grouped into $K$ classes under the assumption that generators in the same class can be treated as identical.

The problem data are as follows:

- $T$ : Number of time steps (h).
- $D_t$: Demand at time period $t$ (MW)
- $R_t$: Generation reserves required at time $t$ (MW)
- $K$ : Set of generator types.
- $G^k$ : Set of generators of type $k$.
- $n_k$ : Number of generators of type $k$
- $b^k_{low}$, $b^k_{high}$ : Coefficients for the piecewise linear approximation of the cost function for a generator of type $k$ (\$/MW).
- $P^k_{low}$, $P^k_{high}$ : Marginal cost coefficients for piecewise linear approximation of the cost function for a generator of type $k$ (\$/MW).
- $\underline{P}^k, \overline{P}^k$ : Minimum and maximum generator limits for generator of type $k$ (MW).
- $RD^k$ : Ramp-down rate of generator of type $k$ (MW/h).
- $RU^k$ : Ramp-up rate of generator of type $k$ (MW/h).
- $SD^k$ : Shutdown limit of generator of type $k$ (MW).
- $SU^k$ : Startup level of generator of type $k$ (MW).
- $UT^k$ : Minimum uptime for generator of type $k$ (h).
- $DT^k$ : Minimum downtime for generator of type $k$ (h).

The variables are:

- $c^k_{t,g}$: Operating cost for generator $g$ of type $k$ at time $t$ (\$).
- $p^k_{t,g}$: Power produced at generator $g$ of type $k$ at time $t$ (MW).
- $u^k_{t,g} \in \{0,1\}$: On/off status of generator $g$ of type $k$ at time $t$.
- $v^k_{t,g} \in \{0,1\}$: Startup status of generator $g$ of type $k$ at time $t$, i.e., whether $g$ is started up at time period $t$.
- $w^k_{t,g} \in \{0,1\}$: Shutdown status of generator $g$ of type $k$ at time $t$, i.e., whether $g$ is shut down at time period $t$.

The problem formulation is as follows:

$$\min \quad \sum_{t=1}^{T} \sum_{k \in K} \sum_{g \in G^k} c^k_{t,g} \tag{8}$$

subject to

$$c^k_{t,g} \geq P^k_{low} p^k_{t,g} + b^k_{low} u^k_{t,g}, \quad \forall g \in G^k, \forall k \in K, t = 1, \ldots, T \tag{9}$$

$$c^k_{t,g} \geq P^k_{high} p^k_{t,g} + b^k_{high} u^k_{t,g}, \ \ \forall g \in G^k, \forall k \in K, t = 1, \ldots, T \tag{10}$$

$$\sum_{k \in K} \sum_{g \in G^k} p^k_{t,g} \geq D_t + R_t, \ \ t = 1, \ldots, T \tag{11}$$

$$\underline{P}^k u^k_{t,g} \leq p^k_{t,g} \leq \overline{P}^k u^k_{t,g}, \ \ \forall g \in G^k, \forall k \in K, t = 1, \ldots, T \tag{12}$$

$$p^k_{t,g} - p^k_{t-1,g} \leq RU^k u^k(t-1) + SU^k v^k_{t,g}, \ \ \forall g \in G^k, \forall k \in K, t = 1, \ldots, T \tag{13}$$

$$p^k_{t-1,g} - p^k_{t,g} \leq RD^k u^k_{t,g} + SD^k w^k_{t,g}, \ \ \forall g \in G^k, \forall k \in K, t = 1, \ldots, T \tag{14}$$

$$\sum_{\ell = t - UT^k + 1, \ \ell \geq 1}^{t} v^k(\ell) \leq u^k_{t,g}, \ \ \forall g \in G^k, \forall k \in K, t = 1, \ldots, T \tag{15}$$

$$u^k_{t,g} + \sum_{\ell = t - DT^k + 1, \ \ell \geq 1}^{t} w^k(\ell) \leq 1, \ \ \forall g \in G^k, \forall k \in K, t = 1, \ldots, T \tag{16}$$

$$u^k_{t-1,g} - u^k_{t,g} + v^k_{t,g} - w^k_{t,g} = 0, \ \ \forall g \in G^k, \ \ \forall k \in K, t = 1, \ldots, T \tag{17}$$

$$c^k_{t,g}, p^k_{t,g} \in \mathbb{R}^+, \ \ \forall g \in G^k, \forall k \in K, t = 1, \ldots, T \tag{18}$$

$$u^k_{t,g}, \ v^k_{t,g}, \ w^k_{t,g} \in \{0, 1\}, \ \ \forall g \in G^k, \forall k \in K, t = 1, \ldots, T \tag{19}$$

The cost function of a generator is generally assumed to be a quadratic function and it is customary to approximate it with a piecewise linear function. This approximation is done here using the two line segments in constraints (9) and (10) that are derived from two tangent lines of the quadratic cost function (representing a low cost and a high cost) strengthened using the method described in [8, 9]. Constraint (11) ensures that enough power is produced to meet demand. Constraints (12) through (17) ensure that each generator's production schedule is feasible [2]. Constraint (12) ensures that each generator's production is within its normal operating limit. Constraints (13) and (14) are ramping constraints ensuring that the output of each generator does not change too rapidly. Constraints (15) and (16) are minimum up and downtime constraints. For these constraints, a negative time index corresponds to a generator's status before the start of the timing horizon. Similarly, a time index larger than $T$ represents the generator's status after the planning horizon has expired. Constraint (17) is a logical constraint, ensuring that the $y^g_t$ variable must take the value of 1 if generator $g$ is turned on at time $t$, and that $z^g_t$ must take the value of 1 if it is turned off. This form of the minimum on and off time constraints was proposed in [25], and tightened ramping constraints were given in [23].

We use the three-binaries-per-generator-hour formulation presented in [2]. This model contains three sets of binary variables at every time period: those representing the on/off status of each generator at the time period, those representing if each generator is started up in the time period, and those representing if each generator is shut down in the time period. Because the startup/shutdown status can be easily determined if the on/off status is known, it is common in the power systems literature to relax the integrality constraints of the startup and shutdown variables.

The binary variables in UC can be expressed using 0/1 matrices. We let $U^k$ be the $T \times n_k$ 0/1 matrix formed using the $u^k_{t,g}$ variables, and we form $C^k$, $P^k$, $V^k$, and $W^k$ similarly. Because any two generators in class $k$ are identical, their production schedules can be permuted to form identical (isomorphic) solutions. Permuting the schedules of two generators of the same type is equivalent to permuting their respective columns in each of $U^k$, $C^k$, $P^k$, $V^k$, and $W^k$. As all generators of the same class are identical, any such permutation of columns will be a symmetry, so long as the permutation is applied to all the matrices. For simplicity in the presentation, we will only consider the variables in $U^k$ that represent the on/off status of each generator of class $k$. We assume that any symmetry that permutes columns of $U^k$ will also permute columns of $C^k$, $P^k$, $V^k$, and $W^k$. This is appropriate for the UC problem because the remaining binary variables can be uniquely determined if the $U^k$ variables are known, and breaking the symmetry of the $U^k$ variables will break all the symmetry between the generators of class $k$.

## 3.2 *Modified Orbital Branching*

We now introduce Modified Orbital Branching (ModOB) as a means to break the symmetry between generators in the same class. We first need some definitions and notation as well as a description of standard Orbital Branching.

### 3.2.1 Symmetry Group and Orbits in MILO

Let $S^n$ denote the set of all permutations of $I^n = \{1, \ldots, n\}$, and let $\mathscr{F}$ denote the feasible set of a given MILO problem on $n$ variables. The *symmetry group* $\mathscr{G}$ of the problem is the set of permutations of the variables that map each feasible solution onto a feasible solution of the same objective function value:

$$\mathscr{G} := \{\pi \in S^n \mid \pi(x) \in \mathscr{F} \text{ and } c^T x = c^T \pi(x) \quad \forall x \in \mathscr{F}\}.$$

Note that computing a problem's symmetry group is NP-hard.

Given the group $\mathscr{G}$, we say that a subset $\{x_{i_1}, x_{i_2}, \ldots, x_{i_k}\}$ of the variables is an *orbit* whenever any of the variables in the set can be mapped to any other using the permutations in $\mathscr{G}$. Therefore the union of the orbits can be interpreted as a partition of the set of variables as well as of the set of indices $I^n$.

### 3.2.2 Orbital Branching

Consider the solution of a 0/1 MILO problem using the standard branch-and-bound technique. A subproblem $a$ in the branch-and-bound tree is defined by two sets: the set of variables fixed to zero, $F_0^a$, the set of variables fixed to one, $F_1^a$. We let $N^a$ be the set of free variables at node $a$. We let $\mathscr{F}^a$ denote the feasible set of $a$ and $\mathscr{G}^a$ its symmetry group. As variables are fixed, $\mathscr{G}^a$ changes and needs to be recomputed. For example, suppose $x_i$ and $x_j$ share an orbit at the root node (there was a permutation in $\mathscr{G}$ that mapped $i$ to $j$). If $x_i \in F_0^a$ and $x_j \in F_1^a$, then for any $x \in \mathscr{F}^a$ and symmetry $\pi$ mapping $i$ to $j$, $\pi(x)_j = 0$, meaning $\pi(x) \notin \mathscr{F}^a$, and thus $\pi$ is not in $\mathscr{G}^a$.

Orbital branching works as follows. Let $\mathscr{O}_i = \{x_{i_1}, x_{i_2}, \ldots, x_{i_k}\}$ be an orbit of $\mathscr{G}^a$. Rather than branching on the disjunction

$$x_{i_1} = 1 \vee x_{i_1} = 0, \tag{20}$$

one uses the branching disjunction

$$\sum_{j=1}^{k} x_{i_j} \geq 1 \vee \sum_{j=1}^{k} x_{i_j} = 0. \tag{21}$$

Note that the right-side disjunction in (21) fixes all the variables involved to zero. While branching decisions typically use disjunctions of type (20), the disjunction (21) is a valid branching decision. Symmetry is exploited by the following observation regarding the left disjunction in (21): since all the variables in $\mathscr{O}_i$ are isomorphic, and at least one of them must be equal to one, any variable in $\mathscr{O}_i$ can be arbitrarily chosen to be one, leading to the symmetry-strengthened disjunction

$$x_{i_1} \geq 1 \vee \sum_{j=1}^{k} x_{i_j} = 0. \tag{22}$$

It is easy to see that (22) is a valid disjunction by considering the following. Because the variables are isomorphic, the subproblem generated by fixing $x_{i_1}$ to one will be equivalent to the subproblem generated by fixing $x_{i_2}$ to one. Therefore, if there is an optimal solution with $x_{i_1}$ equal to one, then there will be an optimal solution with $x_{i_2} = 1$. If there is an optimal solution that is feasible at $a$, there will be an optimal solution feasible in one of the subproblems of $a$.

The strength in orbital branching is that a total of $|\mathscr{O}_i| + 1$ many variables are fixed by the branching, not the two that traditional branching fixes. Because of this, the larger the orbit, the more likely the branching decision will be strong. Stronger branching decisions will improve the lower bound faster, leading to shorter solution times.

On the other hand, the branching trees arising from orbital branching can be significantly unbalanced. This is because the right-hand branch can be much stronger than the left-hand branch as the former fixes $|\mathscr{O}_i|$ variables to 0 while the latter fixes a single variable to 1. Thus the symmetry group of the left subproblem is typically much smaller than that of the current node, and a smaller symmetry group leads to smaller orbits which in turn imply that subsequent branching disjunctions are likely to be weaker. The symmetry group of the right-hand branch does not decrease [22] but this only leads to more powerful branching when pruning does not happen.

### 3.2.3 Modified Orbital Branching

With the objective of achieving a more balanced branch-and-bound tree, suppose that at subproblem $a$ with symmetry group $\mathscr{G}^a$, we branch on orbit $\mathscr{O}_i$ using the following disjunction:

$$\sum_{j=1}^{n} x_{i_j} \geq b' \vee \sum_{j=1}^{n} x_{i_j} \leq b' - 1 \tag{23}$$

for some $b' \in \mathbb{Z}^+$. While different values of $b'$ can be chosen for (23), a natural choice is

$$b' = \lceil \sum_{j \in \mathscr{O}_i} x_j^* \rceil,$$

where $x^*$ is the solution to the LP relaxation of subproblem $a$.

Disjunction (23) can be strengthened in a similar way as (22) strengthens (21). To state this strengthening, we need one more definition. The *projection of the symmetry group* $\Gamma$ on $J \subseteq I^n$ is the set of permutations $\pi_P$ mapping $J$ to $J$ and such that

$$\exists \pi \in \Gamma \text{ s.t. } \pi(x)_j = \pi_P(x)_j \ \forall\, j \in J,\ \forall x \in \mathscr{F}. \tag{24}$$

We denote this set by $\mathtt{Proj}_J(\Gamma)$. Note that it only makes sense to project $\Gamma$ onto a set $J$ if for any $i \in J$, $k$ must be in $J$ if there exists a $\pi \in \Gamma$ with $e_k = \pi(e_i)$. If $J$ represents an orbit of $e_i$ and $\mathtt{Proj}_J(\Gamma)$ consists of all permutations of the elements of the orbit, then $\mathtt{Proj}_J(\Gamma) \cong S^{|J|}$.

The strengthened version of (23) assumes that $\mathtt{Proj}_{\mathscr{O}_i}(\mathscr{G}^a) \cong S^{|\mathscr{O}_i|}$. Under this assumption, if an orbit contains at least $b'$ variables that take the value 1, then $b'$ of the variables can arbitrarily be chosen to take the value 1. Similarly, if at most $b' - 1$ variables take the value 1, then at least $|\mathscr{O}_i| - b' + 1$ variables take the value 0, and under the assumption these variables can be chosen arbitrarily. Thus, if $\mathtt{proj}_{\mathscr{O}_i}(\mathscr{G}^a) \cong S^{|\mathscr{O}_i|}$, the disjunction (23) can be strengthened to

$$x_{i_j} = 1 \; \forall \, j \in \{1, \ldots, b'\} \; \vee \; x_{i_j} = 0 \; \forall \, j \in \{b', b'+1, \ldots, |\mathcal{O}_i|\}. \tag{25}$$

The validity of this strengthening is formally stated in the following theorem.

**Theorem 1 ([24]).** *Let $a = (F_0^a, F_1^a)$ be a node in the branch-and-bound tree with feasible region $\mathcal{F}^a$. Suppose orbit $\mathcal{O}_i$ with $\texttt{proj}_{\mathcal{O}_i}(\mathcal{G}^a) \cong S^{|\mathcal{O}_i|}$ was chosen for branching. Child $\ell$ is formed by fixing $x_{i_j} = 1 \; \forall \, j \in \{1, \ldots, b'\}$ and child $r$ is formed by fixing $x_{i_j} = 0 \; \forall \, j \in \{b', b'+1, \ldots, |O_i|\}$. For any optimal $x^*$ in $\mathcal{F}^a$, there exists a $\pi \in \mathcal{G}^a$ with $\pi(x^*)$ contained in either $\mathcal{F}^\ell$ or $\mathcal{F}^r$.*

We refer the reader to [24] for the proof of Theorem 1.

## 3.3 Modified Orbital Branching and Unit Commitment

In this section, we look at how ModOB can be used to reduce the branch-and-bound search tree when solving the UC problem. The connection between the two is done via the notion of orbitopes.

Recall that the binary variables in UC can be expressed using 0/1 matrices where the rows represent time periods and the columns represent generators, and that for any two generators of the same type $k$, their production schedules can be permuted without changing the optimal value of the solution. We can remove equivalent solutions by restricting the feasible region to matrices with lexicographically non-increasing columns. The convex hull of all $m \times n$ 0/1 matrices with lexicographically non-increasing columns is called a *full orbitope*. This special structure is useful in the context of UC because it makes it possible to efficiently compute the symmetry group for every subproblem in the branch-and-bound tree. Indeed while variable orbits can be computed extremely fast if the symmetry group is known, no polynomial-time algorithm is known for computing the symmetry group of a given subproblem in the context of general integer optimization. By contrast for orbitopes the computation of the column symmetries, and thus of the orbits, can be done in linear time with respect to the number of variables by making use of the following lemma.

**Lemma 1.** *A permutation that permutes columns $c_j$ and $c_j'$ is in the symmetry group of the subproblem $a = (F_0^a, F_1^a) \Leftrightarrow$ either $x_{i,j}$ and $x_{i,j'}$ are fixed to the same value or they are both free for all $i$ in $\{1, \ldots, m\}$.*

Moreover, it can be shown that for a fixed number of time periods in the UC problem, ModOB can enumerate all feasible solutions in polynomial time, and that the branch-and-bound tree grows polynomially as the number of generators increases [24, Sect. 4.1].

Even though ModOB removes a significant proportion of isomorphic solutions from the feasible region, it is still possible for some symmetry to remain unexploited, depending on the branching decisions made. It is possible to remove *all* symmetry using an appropriate branching rule such as the relaxed minimum-rank index (RMRI) branching rule [24, Sect. 4]. We refer the reader to [24] where several such branching rules are proposed and analyzed.

We conclude this section with a summary of the computational results presented in [24] to show the impact of ModOB for the solution of the UC formulation in Section 3.1. The computational experiments were carried out using 25 instances of the UC problem randomly generated based on the generator characteristics described in [3]. The number of generators varied from 46 to 72 and each generator was assigned to one of 8 possible types; each type had between 0 and 19 "identical" generators, depending on the instance. Complete details of the experiments are given in [24].

The results below correspond to the following algorithms:

- **Default CPLEX:** CPLEX's default algorithm. This includes methods such as dynamic search as well as multithreading.
- **B&C (Branch & Cut):** CPLEX with advanced features turned off (mimicking what happens when callbacks are used); CPLEX's symmetry-breaking procedure is used.
- **OB:** Original orbital branching implemented using callbacks.
- **Modified OB:** Modified orbital branching from Section 3.2.3 implemented using callbacks.
- **Modified OB RMRI:** Modified orbital branching implemented using callbacks with the RMRI branching rule to guarantee only non-isomorphic solutions are explored.

All the MILO problems were solved using CPLEX version 12.5.1.0 to within 0.1% optimality, and the number of available threads was set to 1. Nearly all instances were solved within the 2-hour time limit set.

All versions of orbital branching were implemented using the branch callback feature. Branching decisions in the default version are determined by CPLEX, and after CPLEX chooses a branching variable, OB and ModOB compute the orbit of that variable, and branch on that orbit. ModOB+RMRI carries out some additional tests and if necessary branches on a different orbit (see [24] for details). Because callback functions disable other CPLEX features, we also give results for CPLEX's default settings (dynamic search plus additional features) and CPLEX with features disabled (standard branch-and-cut).

The summary of the results is as follows:

| | Computation Time (sec) | | | | |
|---|---|---|---|---|---|
| | CPLEX | B&C | OB | ModOB | ModOB+RMRI |
| Mean | 73.09 | 3523.40 | 1382.86 | 401.39 | 327.70 |
| Geometric Mean | 61.83 | 2183.55 | 830.53 | 273.78 | 243.36 |
| | Number of Nodes | | | | |
| | CPLEX | B&C | OB | ModOB | ModOB+RMRI |
| Mean | 1115.64 | 88378.72 | 26481.44 | 5252.60 | 4222.76 |
| Geometric Mean | 729.25 | 51718.77 | 14359.09 | 2946.22 | 2616.11 |

Let us first compare OB, ModOB and ModOB+RMRI. While ModOB+RMRI gives the best performance among these three variants of orbital branching, the improvement brought about by ModOB in comparison with OB is much more significant than the impact of adding the RMRI branching rule to ModOB. Thus ModOB has a substantial impact on the performance of the MILO solver for this set of UC problems. One possible explanation for this is that the loss of branching flexibility balances out with the full removal of isomorphic solutions. Another possible explanation might be found in the structure of the UC problem. Because of the minimum up and downtimes, branching on one variable can have a significant effect on several other variables. For instance, if a generator is on at time $t$ then off at time $t + 1$, then the generator must be off for several more time periods to satisfy the minimum downtime constraint.

Among all the methods, default CPLEX gives by far the best performance. However, the results show a dramatic difference between CPLEX with its advanced features and CPLEX using callback functions (B&C). One would expect that CPLEX uses a method similar to orbital branching that is seemingly affected by the use of callback functions. Given this large difference, and given the strong impact of ModOB for these instances of UC, it is intriguing to think of how ModOB, even without the RMRI rule, would perform if it were integrated into CPLEX's advanced features.

## 4 Conclusion

Optimization is well established as an essential tool for practitioners in the field of power systems operations. While there have been tremendous improvements in MILO technology since it was first proposed as a means to inform UC decisions more than 50 years ago [10], realistic instances of the UC problem are large-scale and practitioners sometimes still have to settle for solutions that are not globally optimal due to the limited time available to make commitment decisions. Hence the UC problem, like several other fundamental power systems problems, continues to be of interest for researchers. As an example of contemporary research relevant to this area, Section 3 presented Modified Orbital Branching, a new technique to improve the efficiency of MILO solvers when applied to problems that exhibit symmetry. Modified Orbital Branching is particularly relevant when planning the commitment of multiple generators that have identical (or near-identical) operational features.

The advent of the smart grid brings new opportunities for optimization researchers to contribute to the area of power systems. An important example is the need to handle truly enormous sets of data, such as those provided by smart meters, and to unlock the value of these data. In the field of power systems, and more broadly in the area of energy, the impact of Big Data mostly remains to be felt. Another example is the need for customers to actively participate in helping the system operator achieve the necessary balance between supply and demand. This is one of the motivations for the concept of a smart building inside which the various loads are coordinated locally while the system operator (or the utility) influences the decisions on consumption in the building through its pricing of energy. The system architecture summarized in Section 2 provides a means to implement this concept so as to achieve the desired outcome in terms of energy consumption or power modulation. Because the response of single loads will usually not suffice to satisfy the requirements of the grid, effective demand response also requires the aggregation of the coordinated loads' responses in an optimal way. Some recent progress in optimal aggregation of load responses can be found in [13].

In summary, there is an abundance of opportunities for optimization researchers to make major contributions to the practical models and solution algorithms for challenging problems in power systems operations. The author hopes that this chapter will motivate more researchers in optimization to familiarize themselves with the area of power grid operations, and to contribute to the continuing impact of optimization on the operation of this critical infrastructure for our society.

**Acknowledgements** The author thanks the three anonymous referees for their many suggestions for improving this chapter.

## References

1. Anjos, M.F.: Recent progress in modeling unit commitment problems. In: Zuluaga, L.F., Terlaky, T. (eds.) Modeling and Optimization: Theory and Applications: Selected Contributions from the MOPTA 2012 Conference. Springer Proceedings in Mathematics and Statistics, vol. 84. Springer, Berlin (2013)
2. Arroyo, J.M., Conejo, A.J.: Optimal response of a thermal unit to an electricity spot market. IEEE Trans. Power Syst. **15**(3), 1098–1104 (2000)
3. Carrion, M., Arroyo, J.M.: A computationally efficient mixed-integer linear formulation for the thermal unit commitment problem. IEEE Trans. Power Syst. **21**(3), 1371–1378 (2006)
4. Chen, L., Li, N., Jiang, L., Low, S.H.: Optimal demand response: problem formulation and deterministic case. In: Chakarabortty, A., Ilic, M.D. (eds.) Control and Optimization Theory for Electric Smart Grids. Springer, New York (2011)
5. Costanzo, G.T., Kheir, J., Zhu, G.: Peak-load shaving in smart homes via online scheduling. In: IEEE ISIE2011 - International Symposium on Industrial Electronics, Gdansk (2011)
6. Costanzo, G.T., Kosek, A.M., Zhu, G., Ferrarini, L., Anjos, M.F., Savard, G.: An experimental study on load-peak shaving in smart homes by means of online admission control. In: 2012 3rd IEEE PES International Conference and Exhibition on Innovative Smart Grid Technologies (ISGT Europe), pp. 1–8. IEEE, New York (2012)
7. Costanzo, G.T., Zhu, G., Anjos, M.F., Savard, G.: A system architecture for autonomous demand side load management in smart buildings. IEEE Trans. Smart Grid **3**(4), 2157–2165 (2012)
8. Frangioni, A., Gentile, C.: Perspective cuts for a class of convex 0-1 mixed integer programs. Math. Program. **106**, 225–236 (2006)
9. Frangioni, A., Gentile, C., Lacalandra, F.: Tighter approximated milp formulations for unit commitment problems. IEEE Trans. Power Syst. **24**(1), 105–113 (2009)
10. Garver, L.L.: Power generation scheduling by integer programming—development of theory. Trans. Am. Inst. Electr. Eng. Power Apparatus Syst. Part III **81**(3), 730–734 (1962)

11. Gatsis, N., Giannakis, G.B.: Cooperative multi-residence demand response scheduling. In: 2011 45th Annual Conference on Information Sciences and Systems (CISS), pp. 1–6 (2011)
12. Gellings, C.W.: The concept of demand-side management for electric utilities. Proc. IEEE **73**(10), 1468–1470 (1985)
13. Gilbert, F., Anjos, M.F., Marcotte, P., Savard, G.: Optimal design of bilateral contracts for energy procurement. Eur. J. Oper. Res. **246**(2), 641–650 (2015). http://dx.doi.org/10.1016/j.ejor.2015.04.050
14. Kaibel, V., Pfetsch, M.E.: Packing and partitioning orbitopes. Math. Program. **114**, 1–36 (2008)
15. Kaibel, V., Peinhardt, M., Pfetsch, M.E.: Orbitopal fixing. Discret. Optim. **8**(4), 595–610 (2011)
16. Kezunovic, M.: Data integration and information exchange for enhanced control and protection of power systems. In: Proceedings of the 36th Annual Hawaii International Conference on System Sciences. IEEE, New York (2003)
17. Kleissl, J., Agarwal, Y.: Cyber-physical energy systems: focus on smart buildings. In: Proceedings of the 47th Design Automation Conference, DAC '10, pp. 749–754 (2010)
18. Liu, C., Shahidehpour, M., Li, Z., Fotuhi-Firuzabad, M.: Component and mode models for the short-term scheduling of combined-cycle units. IEEE Trans. Power Syst. **24**(2), 976–990 (2009)
19. Margot, F.: Pruning by isomorphism in branch-and-cut. Math. Program. **94**, 71–90 (2002)
20. Margot, F.: Exploiting orbits in symmetric ILP. Math. Program. **98**, 3–21 (2003)
21. Meliopoulos, A.P.S., Cokkinides, G.J., Huang, R., Farantatos, E., Choi, S., Lee, Y., Yu, X.: Smart grid technologies for autonomous operation and control. IEEE Trans. Smart Grid **2**(1), 1–10 (2011)
22. Ostrowski, J., Linderoth, J., Rossi, F., Smriglio, S.: Orbital branching. Math. Program. **126**(1), 147–178 (2009)
23. Ostrowski, J., Anjos, M.F., Vannelli, A.: Tight mixed integer linear programming formulations for the unit commitment problem. IEEE Trans. Power Syst. **27**(1), 39–46 (2012)
24. Ostrowski, J., Anjos, M.F., Vannelli, A.: Modified orbital branching for structured symmetry with an application to unit commitment. Math. Program. **150**(1), 99–129 (2015)
25. Rajan, D., Takriti, S.: Minimum up/down polytopes of the unit commitment problem with start-up costs. Technical report, IBM Research Report (2005)
26. Samadi, P., Mohsenian-Rad, A., Schober, R., Wong, V.W.S., Jatskevich, J.: Optimal real-time pricing algorithm based on utility maximization for smart grid. In: First IEEE International Conference on Smart Grid Communications (SmartGridComm), pp. 415–420 (2010)
27. World Business Council for Sustainable Development. Transforming the market: Energy efficiency in the buildings. Technical report, WBCSD (2009)

# Semidefinite Approaches for MIQCP: Convex Relaxations and Practical Methods

**Hongbo Dong and Nathan Krislock**

**Abstract** We survey several recent advances on applying semidefinite programming (SDP) techniques to globally solve mixed-integer quadratically constrained programs (MIQCPs), or to construct convex relaxations with better tightness/complexity ratios. It is well known that on many MIQCPs, convex relaxations using SDP techniques produce some of the strongest bounds. On one hand, it is commonly thought that SDP relaxations are computationally expensive and hard to exploit in a branch-and-bound framework, where one needs to solve convex relaxations repeatedly. On the other hand, a large amount of effort has been devoted to find more practical ways to exploit strong SDP relaxations. We survey some attempts along this direction, which we conceptually categorized into three main approaches: (1) use first-order methods to solve SDP relaxations approximately; (2) exploit dual optimal solution to derive reformulations; (3) generate cuts in the original variable space by solving the (dual) SDP relaxations heuristically. Our paper does not aim to be a comprehensive survey on MIQCPs, but rather, to be supplementary to other survey papers in literature. We adopt a (partial-) Lagrangian framework in an effort to unify various theoretical and algorithmic developments with (sometimes) simpler notation.

**Keywords** Semidefinite programming • Mixed-integer quadratically constrained programs • Convex relaxations

**MSC (2010):** 90C22, 90C11, 90C26

---

H. Dong (✉)
Department of Mathematics, Washington State University, Pullman, WA, USA
e-mail: hongbo.dong@wsu.edu

N. Krislock
Department of Mathematical Sciences, Northern Illinois University, DeKalb, IL, USA
e-mail: krislock@math.niu.edu



B. Defourny, T. Terlaky (eds.), *Modeling and Optimization: Theory and Applications*, Springer Proceedings in Mathematics & Statistics 147,
DOI 10.1007/978-3-319-23699-5_3

# 1 Introduction

Mixed-integer quadratically constrained program is a large problem class in the following general form

$$\min_{x \in \mathbb{R}^n} \; x^T Qx + q^T x \;\; s.t. \;\; x \in \mathcal{F}, \qquad \text{(MIQCP)}$$

with the feasible region

$$\mathcal{F} := \left\{ x \in \mathbb{R}^n \;\middle|\; \begin{array}{l} x^T Q^{(j)} x + q^{(j)T} x \le h^{(j)}, \forall j \in [m], \\ \ell_i \le x_i \le u_i, \forall i \in [n], \; x_{\mathcal{I}} \in \mathbb{Z}^{|\mathcal{I}|} \end{array} \right\}, \quad \mathcal{I} \subseteq [n],$$

where $[n] = \{1, \ldots, n\}$. The model (MIQCP) is quite general. For example, a polynomial function of $x \in \mathbb{R}^n$ can be reduced to a quadratic function using at most $O(\log(r) \cdot n)$ extra additional variables and additional quadratic constraints, where $r$ is the largest integer in the exponents. The mixed-integer linear program (MILP) is obviously a special case where all quadratic forms are zero. Throughout this paper, we make no assumptions on the convexity of the quadratic objective/constraints, i.e., $Q$ and $Q^{(j)}$ could have positive or negative eigenvalues or both. Therefore there are two sources of nonconvexity in (MIQCP): variable integrality and nonconvex quadratics. In terms of global solution, this could make (MIQCP) more difficult to deal with than MILP, in the sense that its continuous relaxation is a nonconvex QCP, which could be as difficult as (or arguably more difficult than) the original problem.

Compared to the case of MILP, the current state of the art for solving MIQCP to global optimality is rather limited. However, there exist several general purpose software packages, such as BARON [48], Couenne [5], Glomiqo [45] and SCIP [1], that can provide guaranteed global solution on problems up to certain moderate size. For the special case where the quadratic objective and all constraints are convex or second-order-cone representable, some specialized techniques, such as outer-approximation (e.g., [13, 27, 28]) and variants that exploit compact lifted formulations [56], can be used to design more efficient global solution techniques. In this paper we focus on semidefinite programming (SDP) approaches that are applicable to the general (nonconvex) MIQCP case. We refer the readers to the papers [2, 54, 55] for some basics of semidefinite programming.

Many applications of MIQCPs studied in the literature come from combinatorial optimization. The underlying rationale is that in many of these problems, the objective function can be most naturally represented as a quadratic function of the decision variables. One classical example is the Max-Cut problem, where one seeks to partition the nodes of a graph into two sets, while maximizing the total weight of the edges across the partition. In the seminal paper [32], the authors showed that the solution to a semidefinite programming relaxation for the Max-Cut problem can be used to construct feasible solutions with an improved guaranteed approximation ratio. This positive result serves as one motivation for researchers to develop better algorithms to solve semidefinite relaxations (e.g., [8, 18]) as well as global solution strategies based on semidefinite techniques (see [47] for a survey on this line of research). Another classical combinatorial problem that can be formulated as a MIQCP is the Stable-Set problem [41] (or equivalently, the Max-Clique problem). Many researchers also studied MIQCP formulations and semidefinite relaxations for many other problems, such as graph partitioning, quadratic assignment, etc. See [51] for example.

We mention that another large class of applications of MIQCP come from process system optimization. See [44] and the references therein. One of the most important problems in this class is the pooling problem. One survey article on this class of problems is [33].

We remark that our paper is not intended to be a complete survey for MIQCPs. Instead, we focus on the algorithmic efforts of making SDP techniques more practical in a branch-and-bound framework. Our paper can be viewed as supplementary to other survey papers on MIQCP, such as [19] by Burer and Saxena, which focuses more on valid inequalities and constructing tight SDP relaxations.

## 2 SDP relaxations and valid inequalities

The basic semidefinite relaxation for (MIQCP) can be constructed as follows,

$$\begin{aligned} \min_{x\in\mathbb{R}^n, X\in\mathcal{S}^n} \quad & \langle Q, X\rangle + q^T x \\ s.t. \quad & \langle Q^{(j)}, X\rangle + q^{(j)T} x \le h^{(j)}, \quad \forall j, \\ & \ell_i \le x_i \le u_i, \quad \forall i, \\ & X - xx^T \succeq 0. \end{aligned} \tag{SDR}$$

Here $\mathbb{R}^n$ denotes the space of all real vectors of length $n$. $\mathcal{S}^n$ is the space of all $n\times n$ real symmetric matrices. The matrix variable $X$ is introduced to represent the rank-one matrix $xx^T$, and the nonconvex constraint $X = xx^T$ is relaxed to $X - xx^T \succeq 0$. The symbol "$\succeq$" denotes the partial order determined by the positive semidefinite cone in $\mathcal{S}^n$, and $X - xx^T \succeq 0$ means that $X - xx^T$ is positive semidefinite. It is well known that by Schur complement, $X - xx^T \succeq 0$ if and only if

$$\begin{pmatrix} 1 & x^T \\ x & X \end{pmatrix} \succeq 0.$$

Therefore (SDR) is a semidefinite program that can be solved to arbitrary precision by using some interior point algorithms. This relaxation is also called the Shor relaxation [50] due to the fact that it is the relaxation obtained by the technique of Lagrangian relaxation.

The Shor relaxation can be further strengthened by adding convex valid constraints on $x$ and $X$. Most strengthening constraints used in literature are linear, except one case [22, 53] where second-order-cone constraints are used for extended trust-region problems. Valid linear inequalities in $(x, X)$ naturally correspond to quadratic (not necessarily convex) inequalities in $x$. In the following subsections, we briefly mention a few commonly used techniques to strengthen (SDR). Then in our main sections 3 through 5, we discuss practical methods to exploit the resulting SDP relaxation in some global solution framework.

## 2.1 *RLT-type inequalities*

One commonly used technique to generate valid constraints to strengthen (SDR) is the Reformulation-linearization technique (RLT). Provided two linear inequalities that are valid for $\mathcal{F}$, e.g.,

$$a^T x + b \leq 0, \;\; c^T x + d \leq 0 \tag{1}$$

we can multiply them to obtain the valid quadratic inequality

$$(a^T x + b)(c^T x + d) \geq 0, \tag{2}$$

which can be further linearized as

$$\left\langle \frac{ac^T + ca^T}{2}, X \right\rangle + (bc + da)^T x + bd \geq 0. \tag{3}$$

If one of the inequalities in (1) were replaced by an equality, then "$\geq$" in (2) and (3) would also be replaced by "$=$". Many valid constraints used in literature are of this type. For example, if each inequality in (1) simply represents a lower or upper bound of a single variable, then (3) is the well-known McCormick inequality. Note that although the original inequalities (1) may be redundant (or explicitly present) in the original relaxation (SDR), their linearized product (3) can strengthen (sometimes significantly) the semidefinite relaxation. One famous example is the case of BoxQP [3], where $\mathcal{F}$ in (MIQCP) is simply a rectangular region in $\mathbb{R}^n$, and (1) are simply variable bounds. However, it is sometimes non-trivial to find the correct choices of $a, b, c, d$ that are most effective in strengthening (SDR). See, for example, [30] for some discussion on multiplying linear equalities with other linear functions, and [10, 38] for the usage of this technique in different contexts.

We mention the RLT technique can be naturally extended to derive nonlinear constraints of $x$ and $X$. For example, in [22, 53] the authors derived valid second-order-cone constraints to strengthen (SDR) by multiplying a linear inequality and a convex quadratic inequality in $x$.

## 2.2 Copositivity-based inequalities

When $\ell_i \geq 0$, $\forall i$ (note that this can be assumed without loss of generality by either variable shifting or variable-splitting), the positive semidefinite cone constraint $X - xx^T \succeq 0$ can be strengthened using the fact that $x \geq 0$ implies

$$\begin{pmatrix} 1 \\ x \end{pmatrix} \begin{pmatrix} 1 \\ x \end{pmatrix}^T \in \mathcal{CP}_{n+1},$$

where $\mathcal{CP}_k$ is called the *completely positive cone* in the space of $k \times k$ symmetric matrices. A matrix $Y \in \mathcal{CP}_k$ if and only if there exists a matrix $R \in \mathbb{R}^{k \times r}$ such that $Y = RR^T$ and all entries in $R$ are nonnegative. Optimization with the completely positive cone is difficult in general, as it was shown that many NP-hard problems can be formulated exactly as linear program over the completely positive cone. See [14] and the references therein. However, it is possible to exploit partial structure of $\mathcal{CP}_k$ to strengthen the semidefinite relaxation (SDR). First, any matrix in the *dual cone* of $\mathcal{CP}_k$, namely, the *copositive cone*, provides a supporting hyperplane to $\mathcal{CP}_k$, and hence can be used as a valid linear inequality to strengthen (SDR). See [16, 26, 52] for attempts on small scale problems. Second, $\mathcal{CP}_k$ can be relaxed by some of its linear or semidefinite-representable convex relaxations, and several hierarchies of such convex relaxations exist; see, e.g., [12, 24, 37, 46]. One relaxation of $\mathcal{CP}_{n+1}$ that appears to be especially attractive is called the *doubly nonnegative cone*, which is defined as the intersection of the positive semidefinite cone and the cone of elementwise-nonnegative matrices. We will discuss an approach that exploits the structure of the doubly nonnegative cone in Section 3.

## 2.3 Inequalities based on special polytopes

In most MIQCP formulations of combinatorial optimization problems, decision variables are represented by binary variables. One important convex polytope associated with this binary structure is called the *Boolean Quadric Polytope*. For dimension $n$, it is defined as

$$\mathbf{BQP}_n := \mathbf{conv}\left\{(x, y) \in \mathbb{R}^{n+\frac{n(n-1)}{2}} \,\middle|\, x \in \{0,1\}^n,\ y_{ij} = x_i x_j,\ \forall\, 1 \leq i < j \leq n\right\}.$$

Note the $y$ vector corresponds to entries above the diagonal in matrix $X$ in (SDR). Diagonal entries can be strengthened by equalities $X_{ii} = x_i$, by exploiting the quadratic representation of binary variables $x_i \in \{0, 1\}$ if and only if $x_i^2 = x_i$.

In [17], the authors showed that all knowledge of $\mathbf{BQP}_n$ can be transferred to the box-constrained continuous case $x \in [0, 1]^n$, in the sense that $\mathbf{BQP}_n$ is simply a projection of the following set

$$\mathbf{QPB}_n := \mathbf{conv}\left\{(x,X) \in \mathbb{R}^n \times \mathcal{S}^n \middle| x \in [0,1]^n,\ X_{ij} = x_i x_j,\ \forall\ 1 \le i \le j \le n\right\},$$

obtained by dropping the lower-triangular entries of $X$. In other words, any valid inequality for $\mathbf{BQP}_n$ can also be used to strengthen (SDR) provided the feasible set of $x$ is shifted and rescaled into $[0,1]^n$.

$\mathbf{BQP}_n$ is equivalent to another set called the *cut polytope*, in the sense that there exists an affine transformation between $\mathbf{BQP}_n$ and $\mathbf{CUT}_{n+1}$, which is defined as

$$\mathbf{CUT}_n := \mathbf{conv}\left\{y \in \mathbb{R}^{\frac{n(n-1)}{2}} \middle| \exists x \in \{-1,1\}^n,\ y_{ij} = x_i x_j, \forall 1 \le i < j \le n\right\}$$

when using $-1/1$ instead of $0/1$ to represent the binary decisions. Many results on facet-defining valid inequalities for $\mathbf{BQP}_n$ and the cut polytope exist in literature; we refer interested readers to a recent paper [40] and the references therein. We mention that a simple class of facet-defining inequalities for $\mathbf{CUT}_n$ are the *triangle inequalities*, i.e., for $1 \le i < j < k \le n$,

$$\begin{aligned} y_{ij} + y_{ik} + y_{jk} &\ge -1, \\ y_{ij} - y_{ik} - y_{jk} &\ge -1, \\ -y_{ij} + y_{ik} - y_{jk} &\ge -1, \\ -y_{ij} - y_{ik} + y_{jk} &\ge -1. \end{aligned}$$

In subsequent sections, unless otherwise stated, we assume all constraints added to strengthen (SDR) are linear in $x$ and $X$, and can be represented in the following generic form

$$\mathcal{A}(X) + Bx \le b,$$

where $\mathcal{A} : \mathcal{S}^n \mapsto \mathbb{R}^p$ is a linear transformation and $b \in \mathbb{R}^p$.

## 3 Solve semidefinite relaxations by first-order methods

In this and subsequent sections we discuss various practical methods to exploit semidefinite relaxations in globally solving MIQCPs. As interior point methods for SDP are in general computationally demanding and non-trivial to warm-start, they are not among the popular choices for solving (SDR) or its strengthened version in a branch-and-bound algorithm. A natural alternative is to use first-order methods to compute lower bounds by approximately solving semidefinite relaxations. We mention that there are a few general purpose first-order algorithms for semidefinite programming, including the low-rank factorization method [18], the spectral bundle method [36], and the Alternating Direction Augmented Lagrangian method [57], etc. As lower bounding procedures in a branch-and-bound framework, these methods need to be tailored to provide valid lower bounds (even if terminated early) and to warm-start effectively. In this section, we describe two lines of research that use this strategy. Section 3.1 focuses on the special case of all variables being binary, and section 3.2 on the doubly nonnegative relaxations for completely positive formulations of some MIQCPs.

### 3.1 The binary case

In this section we focus on the special case of having only binary variables, i.e., $x_i \in \{0, 1\}, \forall i$. It is well known that the binary condition is equivalent to the quadratic constraint $x_i = x_i^2$, which can be linearized as $x_i = X_{ii}$ to strengthen the semidefinite relaxation (SDR).

Alternatively, a linear transformation (as detailed below) can be used to simultaneously homogenize and transform the binary representation to $-1/1$ representation. There are a few advantages of using this technique. First $-1/1$ representation is used to derive the Geomans-Williamson rounding procedure for the Max-Cut problem in [32]. Second, notation is usually simpler as the linear parts need no special treatment. Third, the rank-1 condition can be equivalently written as one single quadratic constraint on the lifted matrix variable, which we detail in our later discussion on the *BiqCrunch* approach.

Let $U$ denote the $(n+1) \times (n+1)$ invertible matrix $\begin{bmatrix} 1 & 0 \\ -e & 2I \end{bmatrix}$, where $e$ is the all-one vector in proper dimension. We denote

$$Y := U \begin{bmatrix} 1 & x^T \\ x & X \end{bmatrix} U^T.$$

It is then easy to verify that

$$\begin{cases} x_i = X_{ii}, \ \forall i = 1, \ldots, n, \\ \begin{bmatrix} 1 & x^T \\ x & X \end{bmatrix} \succeq 0 \end{cases} \iff \begin{cases} Y_{ii} = 1, \ i = 1, 2, \ldots, n+1, \\ Y \succeq 0. \end{cases} \tag{4}$$

If the matrix variable $Y$ were rank-1 and satisfied (4), then $Y = yy^T$, where $y \in \{-1, 1\}^n$. The semidefinite relaxation (SDR), together with additional strengthening constraints and equalities $x_i = X_{ii}$ for all $i$, can therefore be equivalently reformulated as

$$\begin{aligned} \min_{Y} \ & \langle M, Y \rangle \\ s.t. \ & \langle M^{(j)}, Y \rangle \le h^{(j)}, \quad \forall j \\ & \mathcal{A}(Y) \le b \end{aligned}$$

$$Y_{ii} = 1, \forall i = 1, \ldots, n+1$$
$$Y \succeq 0$$

where $M = U^{-T}\begin{bmatrix} 0 & \frac{1}{2}q^T \\ \frac{1}{2}q^T & Q \end{bmatrix}U^{-1}$, $M^{(j)} = U^{-T}\begin{bmatrix} 0 & \frac{1}{2}q^{(j)T} \\ \frac{1}{2}q^{(j)} & Q^{(j)} \end{bmatrix}U^{-1}$, and $U^{-T} := (U^{-1})^T$.

The main idea in Biq Mac [47] is based on dualizing all lifted linear inequalities $\langle M^{(j)}, Y\rangle \leq h^{(j)}$ and $\mathcal{A}(Y) \leq b$, and then, of those inequalities, to iteratively add the most violated ones. The reason for doing this is because there are typically a very large number of constraints, but only a very small number of these constraints may be active at an optimal solution. For simplicity of notation, we incorporate inequalities $\langle M^{(j)}, Y\rangle \leq h^{(j)}$ into $\mathcal{A}(Y) \leq b$ from now on. Therefore, we now have the semidefinite relaxation

$$\begin{aligned} \min_{Y} \quad & \langle M, Y\rangle \\ s.t. \quad & \mathcal{A}(Y) \leq b \\ & \mathbf{diag}(Y) = e \\ & Y \succeq 0, \end{aligned}$$

where $\mathbf{diag}(Y)$ is the vector along the diagonal of the matrix $Y$ and $e$ is the vector of all ones.

Dualizing only the constraints $\mathcal{A}(Y) \leq b$, we obtain the Langrangian

$$\begin{aligned} L(Y, \mu) &:= \langle M, Y\rangle + \langle \mu, \mathcal{A}(Y) - b\rangle \\ &= \langle M + \mathcal{A}^*\mu, Y\rangle - b^T\mu. \end{aligned}$$

The corresponding dual function is

$$\begin{aligned} \phi(\mu) &:= \min_{Y}\{L(Y, \mu) \mid \mathbf{diag}(Y) = e,\ Y \succeq 0\} \\ &= -b^T\mu + \min_{Y}\{\langle M + \mathcal{A}^*\mu, Y\rangle \mid \mathbf{diag}(Y) = e,\ Y \succeq 0\}, \end{aligned}$$

which is a nonsmooth concave function. The dual of the SDP problem in the above formula is

$$\begin{aligned} \max_{\nu} \quad & e^T\nu \\ s.t. \quad & M + \mathcal{A}^*\mu \succeq \mathbf{Diag}(\nu). \end{aligned}$$

Since both the primal and dual problems are strictly feasible (e.g., one may choose $Y$ to be the identity matrix, and $\nu_i = -t$ for all $i$ where $t$ is sufficiently large), strong duality holds; that is, both primal and dual problems attain their optimal values and there is no duality gap between the primal and dual optimal values. Therefore, we have

$$\phi(\mu) = -b^T\mu + \max_{\nu} \left\{ e^T\nu \mid M + \mathcal{A}^*\mu \succeq \textbf{Diag}(\nu) \right\}.$$

For every $\mu \geq 0$, weak duality implies that $\phi(\mu)$ is a *lower bound* on the optimal value of the original binary quadratic problem. Moreover, if $\mu \geq 0$ and $M + \mathcal{A}^*\mu \succeq \textbf{Diag}(\nu)$, then $-b^T\mu + e^T\nu$ is also a valid lower bound. Such lower bounds can be used to prune the branch-and-bound search tree. The best such lower bound can be obtained by solving the nonsmooth convex optimization problem

$$\max_{\mu \geq 0} \ \phi(\mu).$$

To solve this problem, one needs to evaluate $\phi$ and obtain subgradients of the convex function $-\phi$ for any fixed $\mu$. Evaluating $\phi$ and determining such a subgradient requires solving the primal-dual pair of SDP problems to obtain optimal solutions $Y^*$ and $\nu^*$.

As mentioned before, Biq Mac [47] iteratively adds the most violated inequalities. Initially, none of the inequalities in $\mathcal{A}(Y) \leq b$ are added to the SDP relaxation (hence, $\mu$ is initially the empty vector). Each time the dual function $\phi(\mu)$ is evaluated and an optimal $Y^*$ is obtained, all the inequalities in $\mathcal{A}(Y) \leq b$ are evaluated to determine which inequalities are violated by $Y^*$. A number of the most violated inequalities are then added to the SDP relaxation, and any inequalities that are not active (i.e., the corresponding dual multipliers $\mu_i$ are zero) are removed. This process repeats until the computed lower bound is large enough to prune the branch-and-bound search tree, or until it is determined that it will not be possible to prune the search tree in a reasonable amount of time.

Biq Mac [47] was proved to be very successful at solving many difficult binary quadratic and Max-Cut problems using the strength of semidefinite bounds. Before Biq Mac, many of these problems were not practical to solve using the standard linear programming bounds. Biq Mac was among the first methods to show the practicality of using semidefinite bounds throughout the branch-and-bound search tree. It was reported in [47] that Max-Cut instances with up to 100 variables can be routinely solved, and problems with up to 300 variables having special structure can be solved in a reasonable amount of time. No other algorithm based on LP or convex QP bounding procedure can achieve comparable performance.

We now describe another line of research that built upon the Biq Mac approach and led to the development of the BiqCrunch solver[1]. To further exploit the structure of binary variables, Malick [42] proposed the following valid quadratic *spherical* constraint

$$\|Y\|_F^2 = (n+1)^2,$$

[1] http://lipn.univ-paris13.fr/BiqCrunch/

where $\|Y\|_F$ is the Frobenius norm of the matrix $Y$. Provided $Y \succeq 0$ and $\mathbf{diag}(Y) = e$, this nonconvex quadratic constraint is equivalent to the rank-1 constraint $X = xx^T$. Thus, we have that the problem

$$\begin{aligned} \min_{Y} \quad & \langle M, Y \rangle \\ s.t. \quad & \mathcal{A}(Y) \leq b \\ & \mathbf{diag}(Y) = e \\ & Y \succeq 0 \\ & \|Y\|_F^2 = (n+1)^2 \end{aligned}$$

is equivalent to the original binary quadratic programming problem. Relaxing the hard spherical constraint by dualizing it (see [43]), one obtains the following regularized semidefinite relaxation

$$\begin{aligned} \min_{Y} \quad & \langle M, Y \rangle + \frac{\alpha}{2}\left(\|Y\|_F^2 - (n+1)^2\right) \\ s.t. \quad & \mathcal{A}(Y) \leq b \\ & \mathbf{diag}(Y) = e \\ & Y \succeq 0, \end{aligned}$$

where $\alpha$ is a nonnegative scalar. Since $\|Y\|_F^2 \leq (n+1)^2$ for all $Y \succeq 0$ having $\mathbf{diag}(Y) = e$, we have a possibly weaker bound for $\alpha > 0$.

Proceeding by forming the Langrangian of the relaxed problem, we have

$$\begin{aligned} L_\alpha(Y, \mu, \nu) &:= \langle M, Y \rangle + \frac{\alpha}{2}\left(\|Y\|_F^2 - (n+1)^2\right) + \langle \mu, \mathcal{A}(Y) - b \rangle + \langle \nu, e - \mathbf{diag}(Y) \rangle \\ &= \langle M + \mathcal{A}^*\mu - \mathbf{Diag}(\nu), Y \rangle + \frac{\alpha}{2}\|Y\|_F^2 - b^T\mu + e^T\nu - \frac{\alpha}{2}(n+1)^2. \end{aligned}$$

We then obtain the corresponding dual function

$$\begin{aligned} \phi_\alpha(\mu, \nu) &:= \min_{Y \succeq 0} L_\alpha(Y, \mu, \nu) \\ &= -\frac{1}{2\alpha}\|[M + \mathcal{A}^*\mu - \mathbf{Diag}(\nu)]_-\|_F^2 - b^T\mu + e^T\nu - \frac{\alpha}{2}(n+1)^2, \end{aligned}$$

where $[M + \mathcal{A}^*\mu - \mathbf{Diag}(\nu)]_-$ is the negative definite matrix that is closest to the matrix $M + \mathcal{A}^*\mu - \mathbf{Diag}(\nu)$ with respect to the Frobenius norm—for details on the above approach for the Max-Cut problem, see [39]. The distinguishing feature of this dual function is that it is not only concave, but it is also a *smooth* function with known gradient. In addition, to evaluate the dual function $\phi_\alpha(\mu, \nu)$ and its gradient, one needs only to compute the negative eigenvalues and corresponding eigenvectors of the matrix $M + \mathcal{A}^*\mu - \mathbf{Diag}(\nu)$, so no SDP problem need be solved in this case.

Again, by weak duality, for any $\mu \geq 0$ and $\nu$, the value of $\phi_\alpha(\mu, \nu)$ is a *lower bound* on the optimal value of the original binary quadratic problem. However, the best such lower bound can now be obtained by solving the *smooth* convex optimization problem

$$\max_{\mu \geq 0,\ \nu} \ \phi_\alpha(\mu, \nu).$$

As with the Biq Mac method, the method appearing in [39] (and also used in BiqCrunch) iteratively adds the most violated inequalities. For a fixed value of $\alpha$, the function $\phi_\alpha(\mu, \nu)$ is maximized using a quasi-Newton method, returning a matrix $Y_\alpha$ that is related to the gradient of $\phi_\alpha(\mu, \nu)$. The matrix $Y_\alpha$ is then used to determine which inequalities are violated—a number of the most violated inequalities are then added, and any inequalities that are not active are removed. Once there are only a small number of inequalities that are violated by a fixed level, the value of $\alpha$ is reduced. The procedure is repeated until the lower bound is large enough to prune the branch-and-bound search tree or until it is determined that it will not be possible to prune in a reasonable amount of time.

It was demonstrated in [39] that the above regularized approach generated SDP-quality bounds in less time than was required by the bounding procedure of Biq Mac. Moreover, the branch-and-bound method proved to be more robust at solving difficult binary quadratic and Max-Cut problems. BiqCrunch is an implementation of this improved bounding procedure for general binary quadratic problems with arbitrary quadratic constraints, and the open-source code is available for download from the BiqCrunch website.

## 3.2 *The case of doubly nonnegative (DNN) programs*

Motivated by a result that a class of MIQCP can be equivalently formulated as linear programs over the completely positive cone [21], Sam Burer in [15] proposed an alternating projection augmented Lagrangian algorithm to solve the doubly nonnegative (DNN) relaxation for completely positive formulations of MIQCP in [21]. Combined with an idea of finite branching with complementarity conditions [20], Burer and Chen [23] implemented a branch-and-bound algorithm for globally solving nonconvex quadratic programming with linear constraints and continuous variables.

The procedure of deriving DNN relaxation can be alternatively described by the following three steps;

1. Use standard techniques, e.g., adding slack variables, variable shifting and splitting, to reformulate the original MIQCP so that all linear conditions (i.e., linear inequalities, variable upper bounds) are represented by linear equations $Ax = b$ and nonnegativity of all variables $x \in \mathbb{R}^n_+$; further we require that all variables are either binary or continuous;

2. Derive the basic semidefinite relaxation (SDR);
3. Strengthen (SDR) by the RLT-type valid constraints

$$\begin{aligned} X_{ii} &= x_i, \ \forall x_i \ binary \\ \mathbf{diag}(AXA^T) &= b \circ b, \\ X_{ij} &\geq 0, \ \forall i,j \end{aligned}$$

where $b \circ b$ is the vector with entries $b_j^2$, $\forall j$.

The algorithm in [15] applies to the case that all other quadratic constraints in (MIQCP) are in the form of $x_i x_j = 0 (\forall (i,j) \in E \subseteq [n]^2)$. The basic idea of their algorithm is to create two duplicated copies of the matrix variable $X$, say, $Y$ and $Z$. All "semidefinite" constraints in the DNN relaxation are then written in terms of $Y$ and all "linear" constraints are written in terms of $Z$. The "linking" constraint $Y = Z$ is then relaxed and penalized by an augmented Lagrangian term. Let the dual variable associated with the linking constraint be $S$, their algorithm performs alternating updates on $Y$, $Z$, and $S$. The computation of each update is relatively simple, i.e., via an eigenvalue factorization or some computation of linear complexity. Their computational results showed that this approach is quite effective in solving several classes of problems to global optimality, including BoxQP and the quadratic multiple knapsack problem. Further, Burer and Chen [23] implemented a global solver for nonconvex quadratic programming, namely *quadprogbb* (i.e., a global version of the Matlab function *quadprog*) based on the idea of this development.

In [4], the authors established some theoretical dominance relations between various semidefinite relaxations for Quadratically Constrained Programs with *continuous* variables. However, note that this is sufficiently general, as any (bounded) variable integrality can be represented using a group of binary variables, while each binary variable can be represented as a continuous variable with an additional quadratic constraint. The authors showed that the DNN relaxation is equivalent to the classical Shor relaxation strengthened by RLT inequalities exploiting all variable bounds and linear constraints. Finally, they showed that the DNN relaxation is, theoretically and computationally, among the strongest semidefinite relaxations in their comparison.

## 4 Exploit the dual optimal solution to SDP relaxations

We now turn to describe a second type of practical approaches to exploit semidefinite relaxations for MIQCP, which are usually called the *Quadratic Convex Reformulation* (QCR) methods. These methods aim to derive a convex-mixed-integer reformulation for the original MIQCP in a pre-processing step. The resulting reformulation has the advantage that its continuous relaxation is a convex program, therefore easier to deal with in a branch-and-bound algorithm. We attempt to adopt a general approach that is independent of the structure of the feasible region $\mathcal{F}$ whenever possible. In Section 4.1 we show how to construct convex quadratic programming (QP) relaxations by using the dual solution to an SDP relaxation. Provided strong duality and dual attainment of the SDP, this convex QP provides the same lower bound as the SDP relaxation. In Section 4.2 we show how additional linear variables can be incorporated into this procedure. In Section 4.3 we show how a convex QP *reformulation* for the original MIQCP can be obtained, and how different studies in the literature fit into our framework.

### 4.1 Convex QP that provides the same bound as the SDP relaxation

We start with the general formulation (MIQCP). The general QCR method [11] could be realized as the following procedure. Provided a semidefinite relaxation

$$\begin{aligned} p^* = \min_{x,X} \ & \langle Q, X\rangle + q^T x \\ s.t. \ & \mathcal{A}(X) + Bx \le b \\ & X - xx^T \succeq 0 \end{aligned} \tag{PSDR}$$

which could be (SDR) or its strengthened version, we obtain the dual problem for (PSDR)

$$\begin{aligned} d^* = \max_{y\ge 0,t} \ & -b^T y - t \\ s.t. \ & \begin{pmatrix} t & \frac{1}{2}(B^T y + q)^T \\ \frac{1}{2}(B^T y + q) & Q + \mathcal{A}^* y \end{pmatrix} \succeq 0. \end{aligned} \tag{DSDR}$$

For the simplicity of presentation, we assume that strong duality holds for this dual pair of SDPs, and that the primal and dual optimal values are attained at $(\bar{x}, \bar{X})$ and $(\bar{t}, \bar{y})$, respectively. Then, by dualizing $\mathcal{A}(X) + Bx \le b$, a further relaxed SDP is

$$p^* \ge \zeta^* = \min_{x,X} \langle Q, X\rangle + q^T x + \bar{y}^T \left(\mathcal{A}(X) + Bx - b\right) \ \ s.t. \ X - xx^T \succeq 0. \tag{5}$$

Because $Q + \mathcal{A}^*(\bar{y}) \succeq 0$, we may assume $X = xx^T$ without loss of generality; this is because if $X \succeq xx^T$, then

$$\begin{aligned} \langle Q, X\rangle + q^T x + \bar{y}^T \left(\mathcal{A}(X) + Bx - b\right) &= \langle Q + \mathcal{A}^* \bar{y}, X\rangle + (B^T \bar{y} + q)^T x - b^T \bar{y} \\ &\ge \langle Q + \mathcal{A}^* \bar{y}, xx^T\rangle + (B^T \bar{y} + q)^T x - b^T \bar{y}, \end{aligned}$$

so replacing $X$ with $xx^T$ results in an objective value that is at least as good. Therefore the SDP in equation (5) is equivalent to the following unconstrained convex quadratic program

$$\zeta^* = \min_x \; x^T \left(Q + \mathcal{A}^* \bar{y}\right) x + (B^T \bar{y} + q)^T x - b^T \bar{y}. \tag{6}$$

Furthermore, we have $\zeta^* \geq d^*$ since

$$\begin{aligned} 0 &\leq \begin{pmatrix} 1 \\ x \end{pmatrix}^T \begin{pmatrix} \bar{t} & \frac{1}{2}(B^T \bar{y} + q)^T \\ \frac{1}{2}(B^T \bar{y} + q) & Q + \mathcal{A}^* \bar{y} \end{pmatrix} \begin{pmatrix} 1 \\ x \end{pmatrix} \\ &\Rightarrow x^T \left(Q + \mathcal{A}^* \bar{y}\right) x + (B^T \bar{y} + q)^T x - b^T \bar{y} \geq -b^T \bar{y} - \bar{t}, \end{aligned}$$

for all $x$. Therefore, problem (6) provides the same bound as (PSDR) if $p^* = d^*$.

One might think that adding further constraints to (6), say $x \in \mathcal{R}$, which is a superset of $\mathcal{F}$ - the feasible region of (MIQCP) - may strengthen the bound. However, this is not the case unless $\mathcal{R}$ includes some new "relaxation strength" other than the feasible region (PSDR). In other words, if

$$\left\{ x \,\middle|\, \exists X, \begin{array}{r} \mathcal{A}(X) + Bx \leq b \\ X - xx^T \succeq 0 \end{array} \right\} \subseteq \mathcal{R}, \tag{7}$$

then adding $x \in \mathcal{R}$ to (6) has no effect on the bound. This can be proved by showing that, exploiting the optimality condition of (PSDR) and (DSDR), the optimal value of (6) is attained at $\bar{x}$, which is part of the optimal solution to $(PSDR)$, and that $\bar{x} \in \mathcal{R}$ given the condition (7).

**"Primalize" simple constraints.** Provided that $\mathcal{A}(X) + Bx \leq b$ includes some constraints that do not involve $X$, say in the form of $Ex \leq f$, and further let $\bar{y}$ be optimal in (DSDR), and $\bar{\bar{y}}$ be the modification of $\bar{y}$ such that the dual variables corresponding to $Ex \leq f$ are set to zero, then the following constrained convex QP provides the same bound as (6),

$$(p^* = \zeta^* = d^* =)\bar{\zeta}^* := \min_x \; x^T \left(Q + \mathcal{A}^* \bar{\bar{y}}\right) x + (B^T \bar{\bar{y}} + q)^T x - b^T \bar{\bar{y}}, \;\; s.t. \;\; Ex \leq f. \tag{8}$$

To show the equality of the bounds, first note (8) is again a (partial) Lagrangian relaxation of (PSDR), hence $p^* \geq \bar{\zeta}^*$. Next, since $\bar{y} \geq 0$, for any $x$ such that $Ex \leq f$, we have

$$\bar{\bar{y}}^T (\mathcal{A}(xx^T) + B^T x - b) \geq \bar{y}^T (\mathcal{A}(xx^T) + B^T x - b),$$

and the objective function in (8) is no less than that in (6). Therefore $\bar{\zeta}^* \geq \zeta^* = d^*$. This idea is useful in our later construction of *convex-mixed-integer reformulations* (section 4.3) and is implicitly used in some of the studies mentioned there.

To sum up, we have the following theorem.

**Theorem 1.** *Let (PSDR) and (DSDR) be the primal-dual pair of a semidefinite relaxation of (MIQCP). Suppose that strong duality holds for (PSDR) and (DSDR), and primal and dual optimal solutions are attained at* $(\bar{x}, \bar{X})$ *and* $(\bar{t}, \bar{y})$*, respectively. Then (6) is an unconstrained convex QP relaxation of (MIQCP) providing the same bound, i.e.,* $p^* = \zeta^* = d^*$*. Further, if the constraints* $\mathcal{A}(X) + Bx \leq b$ *include linear constraints* $Ex \leq f$*, and* $\bar{\bar{y}}$ *is a modification of* $\bar{y}$ *by setting the dual variables corresponding to* $Ex \leq f$ *to be zero, then (8) is a constrained QP relaxation of (MIQCP) providing the same bound, i.e.,* $p^* = \zeta^* = \bar{\zeta}^* = d^*$.

## 4.2 Using additional variables

In many cases, tighter relaxations may be derived by using additional lifted variables. Consider the following SDP relaxation using an additional vector of variables $z$.

$$\begin{aligned} p^* = \min_{x,X} \ & \langle Q, X\rangle + q^T x \\ s.t. \ & \mathcal{A}(X) + Bx + Cz \leq b \\ & X - xx^T \succeq 0 \end{aligned} \tag{PSDR\_z}$$

The corresponding dual is

$$\begin{aligned} d^* = \max_{y \geq 0, t} \ & -b^T y - t \\ s.t. \ & C^T y = 0 \\ & \begin{pmatrix} t & \frac{1}{2}(B^T y + q)^T \\ \frac{1}{2}(B^T y + q) & Q + \mathcal{A}^* y \end{pmatrix} \succeq 0. \end{aligned} \tag{DSDR\_z}$$

Again we assume that strong duality holds and both primal and dual optimal values are attained. By a similar argument as in the previous section, we obtain a convex QP relaxation with the same bound,

$$\zeta^* = \min_{x,z} \ x^T \left(Q + \mathcal{A}^* \bar{y}\right) x + (B^T \bar{y} + q)^T x + (C^T \bar{y})^T z - b^T \bar{y}. \tag{9}$$

One can similarly show that $p^* \geq \zeta^* \geq d^*$, and all three values coincide when $p^* = d^*$. The argument of "primalize" linear constraints of $x$ and $z$ follows similarly.

In [11], the authors used a $z$ vector[2] to additionally (other than $X$) represent the vector $\mathbf{vec}(xx^T)$. Here $\mathbf{vec}(\cdot)$ denotes the long vector obtained by stacking the

---

[2]In their notation this vector is called $y$, see problem $(QP_{\alpha,\beta})$ in Section 3 of [11]

columns one after another. In their case, $\mathcal{A}(X) + Bx + Cz \leq b$ contains linear equalities $\mathbf{vec}(X) = z$ and the McCormick inequalities using $x$ and $z$. In their version of the primal SDP relaxation (PSDR_z) (i.e., Theorem 2 in Section 3.2 of [11]), the $z$ vector[3] does not appear because it can be replaced by entries in $X$. The main reason that they introduced this redundant version of $X$ is that they can (by adding additional constraints to (9)) enforce $\bar{y}^T\left(\mathcal{A}(xx^T) + Bx + Cz - b\right) = 0$ for all $x \in \mathcal{F}$, provided that $x \in \mathcal{F}$ implies $x$ is a integral vector. This equality is needed to construct a convex-mixed-integer *reformulation* of (MIQCP) by combining (9) and some additional constraints. We will discuss convex-mixed-integer reformulations in the next subsection.

## 4.3 Constructing convex-mixed-integer reformulations

Suppose that $(\bar{t}, \bar{y})$ is feasible in (DSDR), by construction we have $\bar{y}^T(\mathcal{A}(xx^T) + Bx - b) \leq 0$ for all $x \in \mathcal{F}$. The unconstrained QP relaxation (6) could be further strengthened by adding the original (nonconvex) constraint $x \in \mathcal{F}$,

$$\min_x \; x^T\left(Q + \mathcal{A}^*\bar{y}\right)x + (B^T\bar{y} + q)^T x - b^T\bar{y}, \quad s.t. \quad x \in \mathcal{F}. \tag{10}$$

If the SDP relaxation (PSDR) is constructed such that

$$\bar{y}^T\left(\mathcal{A}(xx^T) + Bx - b\right) = 0, \; \forall x \in \mathcal{F}, \tag{11}$$

then (10) is a *reformulation* of the original problem (MIQCP), in the sense that every feasible solution in (PSDR) is feasible in (10) with the same objective value, and vice versa. We remark that in many cases, (11) is satisfied by constructing (PSDR) such that

$$\mathcal{A}(xx^T) + Bx - b = 0, \forall x \in \mathcal{F}. \tag{12}$$

Similarly for the case where additional variables are used, as in section 4.2, as long as the semidefinite relaxation (PSDR_z) is constructed such that

$$\bar{y}^T\left(\mathcal{A}(xx^T) + Bx + Cz - b\right) = 0, \; \forall (x, z) \in \tilde{\mathcal{F}}, \tag{13}$$

for some dual feasible solution $(\bar{t}, \bar{y})$, a reformulation of (MIQCP) is

$$\min_x \; x^T\left(Q + \mathcal{A}^*\bar{y}\right)x + (B^T\bar{y} + q)^T x + (C^T\bar{y})^T z - b^T\bar{y}, \quad s.t. \quad (x, z) \in \tilde{\mathcal{F}}, \tag{14}$$

where $\tilde{\mathcal{F}}$ is a set in the lifted space such that the projection onto the $x$ variables is exactly $\mathcal{F}$.

---

[3] Again, $y$ in their notation.

Compared to (MIQCP), (10) or (14) has the advantage that they have convex objective functions. They are especially easier to deal with when $\mathcal{F}$ can be represented as a polyhedral set with mixed-integer constraints. In this case (10) can be solved efficiently by using some efficient implementations of branch-and-bound algorithms, such as those used in software packages, Gurobi [34], etc. Further, the root bound of (10) cannot be worse than the bound of (PSDR) or (PSDR_z) when strong duality holds. However, in many cases, as we discuss below, the continuous relaxation of $\mathcal{F}$ satisfies the condition in (7), and then the root bound of (10) is the same as the SDP bound.

We now relate the general framework presented here to various special cases studied in the literature. Those studies mainly differ in their way to choose constraints $\mathcal{A}(X) + Bx \leq b$ in (PSDR), or $\mathcal{A}(X) + Bx + Cz \leq b$ in (PSDR_z).

For the simplest case of $\mathcal{F} = \{0, 1\}^n$, it is well known that the following equalities are valid,

$$X_{ii} - x_i = 0, \forall i, \ x_i \in \{0, 1\}. \tag{15}$$

The convexification idea in an early paper [35] amounts to choosing (PSDR) with $\mathcal{A}(X) + Bx \leq b$ as $\mathbf{tr}(X) - \sum_{i=1}^{n} x_i = 0$, which is an aggregation of (15). In [9] the authors used all equalities (15) in $\mathcal{A}(X) + Bx \leq b$, hence added more flexibility in the penalization term $\bar{y}^T \left(\mathcal{A}(xx^T) + Bx - b\right)$, which is a linear combination of expressions $x_i^2 - x_i, \ \forall i$.

When $\mathcal{F}$ further includes a set of linear equality constraints, i.e.,

$$\mathcal{F} \subseteq \{x \in \mathbb{R}^n | a_j^T x = b_j, j = 1, \ldots, m\} \cap \{0, 1\}^n,$$

further constraints can be added into $\mathcal{A}(X) + Bx \leq b$ by exploiting these equalities while maintaining (12). In [10], the authors further used equalities

$$\begin{pmatrix} 1 & x^T \\ x & X \end{pmatrix} \begin{pmatrix} -b_j \\ a_j \end{pmatrix} = 0, \ \ j = 1, \ldots, m. \tag{16}$$

For each $j$, the last $n$ rows are exactly the RLT-type quadratic equalities obtained by multiplying each linear equality with a single variable, i.e., $x_i(a_j^T x - b_j) = 0, \forall i = 1, \ldots, n, j = 1, \ldots, m$. By using a result in [30], the authors of [10] also showed that all the $m \cdot n$ constraints in (16), when added to (PSDR), are equivalent to the following *single* linear equality

$$\sum_{j=1}^{m} \begin{pmatrix} -b_j \\ a_j \end{pmatrix}^T \begin{pmatrix} 1 & x^T \\ x & X \end{pmatrix} \begin{pmatrix} -b_j \\ a_j \end{pmatrix} = 0. \tag{17}$$

We include a simple proof here for the sake of completeness.

**Lemma 1 ([21, 30]).** *Suppose that* $\begin{pmatrix} 1 & x^T \\ x & X \end{pmatrix} \succeq 0$, *then (17) holds if and only if (16) holds.*

*Proof.* The proof of (16) implying (17) is straightforward. Let $\begin{pmatrix} 1 & x^T \\ x & X \end{pmatrix} = UU^T$, then (17) implies

$$\sum_{j=1}^{m} \left\| U^T \begin{pmatrix} -b_j \\ a_j \end{pmatrix} \right\|_2^2 = 0 \quad \Rightarrow \quad U^T \begin{pmatrix} -b_j \\ a_j \end{pmatrix} = 0, \forall j$$

$$\Rightarrow \quad UU^T \begin{pmatrix} -b_j \\ a_j \end{pmatrix} = 0, \forall j \quad \Rightarrow \quad (16). \qquad \square$$

In [11], the authors extended the QCR method to the following case where $\mathcal{F}$ can be represented as a polyhedral set $\mathcal{P}$ with general mixed integer constraints:

$$\mathcal{F} = \mathcal{P} \cap \left\{ x = [x_{\mathcal{I}}; x_{\mathcal{J}}] \in \mathbb{Z}^{|\mathcal{I}|} \times \mathbb{R}^{|\mathcal{J}|} \,\middle|\, 0 \le x_i \le u_i, \ \forall i \in \mathcal{I} \cup \mathcal{J} \right\}.$$

First of all, by adding slack variables, the authors converted all inequalities into equalities and assumed that $\mathcal{P} = \{x \in \mathbb{R}^n | a_j^T x = b_j, j = 1, \ldots, m\}$. Note that merely exploiting the equalities by adding (17) into (PSDR) may not be sufficient to derive a convex-mixed-integer reformulation, as $Q + \mathcal{A}^* y$ may never be positive semidefinite no matter what the choice of $y$ is. The authors further strengthened (PSDR) by exploiting the following McCormick inequalities between variables

$$\max(0, u_j x_i + u_i x_j - u_i u_j) \le x_i x_j \le \min(u_j x_i, u_i x_j), \quad \forall (i,j), i \in \mathcal{I} \text{ or } j \in \mathcal{I}. \qquad (18)$$

Unfortunately, if one simply replaces $x_i x_j$ with $X_{ij}$ and incorporates into $\mathcal{A}(X) + Bx \le b$ in (PSDR), again one may not obtain a reformulation since (11) will not hold in general. To overcome this difficulty, the authors used the framework illustrated in section 4.2, and introduced vector $z$ (denoted by $y$ in [11]) as another "linearized version" of entries in $\mathbf{vec}(xx^T)$. Essentially the authors added equalities

$$X_{ij} - z_{ij} = 0, \quad \forall (i,j), i \in \mathcal{I} \text{ or } j \in \mathcal{I}, \qquad (19)$$

and (18)—linearized by replacing $x_i x_j$ with the corresponding variable in $z$—into $\mathcal{A}(X) + Bx + Cz \le b$. Strong duality and dual attainment hold for the resulting SDP pair (PSDR_z) and (DSDR_z).

In order to construct a convex-mixed-integer reformulation, first the authors implicitly used the idea of "primalize" simple constraints, as explained in section 4.1. Denote the modified dual variable to be $\hat{y}$. It then suffices to ensure that $\hat{y}^T \left( \mathcal{A}(xx^T) + Bx - b \right) = 0$ for all $x \in \mathcal{F}$. The main work is to ensure that

$$x_i x_j - z_{ij} = 0, \quad \forall (i,j), i \in \mathcal{I} \text{ or } j \in \mathcal{I}$$

hold when integrality conditions are satisfied in the relaxed problem. The authors constructed a lifted polyhedral set with mixed-binary conditions, $(x, z, t) \in \tilde{\mathcal{F}}$, where $t$ is some additional lifted variables, such that the projection onto $x$ equals $\mathcal{F}$. The main techniques used to construct $\tilde{\mathcal{F}}$ are the binary expansion of an integer variable $x_i(\forall i \in \mathcal{I})$and the convex hull representation of triplets $(t_i, t_j, t_i \cdot t_j)$ when at least one of $t_i$ and $t_j$ is binary.

In [31] the authors consider the case where $\mathcal{F}$ is a quadratically constrained set with binary variables, i.e.,

$$\mathcal{F} := \left\{x \in \{0,1\}^n \middle| x^T Q^{(i)} x + q^{(i)T} x \le h_i, i = 1, \ldots, r\right\}.$$

Note that $\mathcal{F}$ can still be represented as a convex-mixed-integer set. For example, each quadratic form in $\mathcal{F}$ can be perturbed by using a linear combination of $x_i^2 - x_i, \forall i$ such that all quadratic inequalities become convex, while keeping the constraint unchanged when $x$ is integral. We denote the representation of $\mathcal{F}$ obtained in this way as $\mathcal{F}_{cvx}$. We assume that in (PSDR), $\mathcal{A}(X) + Bx \le b$ contains equalities $\mathbf{diag}(X) - x = 0$ and inequalites $\langle Q^{(i)}, X\rangle + q^{(i)T} x \le h_i, \forall i$. Then by the argument in section 4.1, problem (10)—with $\bar{y}$ chosen as an optimal solution to (DSDR) and $\mathcal{F}$ replaced by $\mathcal{F}_{cvx}$—is a mixed-integer-convex *relaxation* (not yet a reformulation) whose continuous relaxation bound equals the SDP bound (PSDR).

Extra work is needed to obtain a reformulation. The authors of [31] provided a simple solution by exploiting the fact that all variables are binary. First, as $\bar{y}$ is the optimal dual solution to (DSDR), the following SDP relaxation provides the same bound as (PSDR),

$$\begin{aligned} p^* = \min_x \quad & \langle Q, X\rangle + q^T x, \\ s.t. \quad & \bar{y}^T(\mathcal{A}(X) + Bx - b) + s = 0, \\ & X - xx^T \succeq 0, \ s \ge 0. \end{aligned}$$

Its dual problem is the following uni-variate problem

$$\begin{aligned} \max_{\epsilon \ge 0} \quad & -(b^T\bar{y}) \cdot \epsilon - t \\ s.t. \quad & \begin{pmatrix} t & \frac{1}{2}(\epsilon B^T\bar{y} + q)^T \\ \frac{1}{2}(\epsilon B^T\bar{y} + q) & Q + \epsilon \mathcal{A}^*\bar{y} \end{pmatrix} \succeq 0. \end{aligned} \qquad (DSDR_\epsilon)$$

By comparing $(DSDR_\epsilon)$ with (DSDR), and using a simple change-of-variable argument, one can see that the optimal value of $(DSDR_\epsilon)$ equals $d^*$—the optimal value of (DSDR)—and can be attained at $\epsilon = 1$. By an argument similar to section 4.1, the following convex QP has the same relaxation bound as (SDR),

$$p^* = \min_{x,s} \; x^T \left(Q + \mathcal{A}^*\bar{y}\right) x + (B^T\bar{y} + q)^T x - b^T\bar{y} + s, \tag{20}$$
$$\text{s.t.} \;\; s \geq 0.$$

Therefore a reformulation of (MIQCP) can be obtained by explicitly enforcing the condition that the perturbation term, $\bar{y}^T(\mathcal{A}(xx^T) + Bx - b) + s$ , must be exactly zero, i.e.,

$$\begin{aligned} \min_{x,s} \;\; & x^T \left(Q + \mathcal{A}^*\bar{y}\right) x + (B^T\bar{y} + q)^T x - b^T\bar{y} + s \\ \text{s.t.} \;\; & \bar{y}^T(\mathcal{A}(xx^T) + Bx - b) + s = 0, \\ & s \geq 0, x \in \mathcal{F}_{cvx}. \end{aligned}$$

To further construct a convex-mixed-integer reformulation, note that the nonconvex equality can be split into two inequalities, where each of them can be convexified by adding proper linear combinations of expressions $x_i^2 - x_i, \forall i$. For example, one may choose $\lambda_i$ and $\mu_i$ large enough such that the following problem is a mixed-integer-convex program,

$$\begin{aligned} \min_{x,s} \;\; & x^T \left(Q + \mathcal{A}^*\bar{y}\right) x + (B^T\bar{y} + q)^T x - b^T\bar{y} + s, \\ \text{s.t.} \;\; & \bar{y}^T(\mathcal{A}(xx^T) + Bx - b) + s + \sum_i \lambda_i(x_i^2 - x_i) \leq 0 \\ & -\bar{y}^T(\mathcal{A}(xx^T) + Bx - b) - s + \sum_i \mu_i(x_i^2 - x_i) \leq 0 \\ & s \geq 0, x \in \mathcal{F}_{cvx}. \end{aligned} \tag{21}$$

Then for integral $x$, the equality constraint $\bar{y}^T(\mathcal{A}(xx^T) + Bx - b) + s = 0$ must hold, which implies the objective function in (21) equals the true objective function $x^TQx + q^Tx$. Therefore (21) is indeed a mixed-integer-convex reformulation to (PSDR).

We finally give some remarks on the computational performance of solving the convex-mixed-integer reformulations presented in this section. For the case of binary quadratic programming, or equivalently, the Max-Cut problem, it has been show in [9] that QCR is advantageous over a mixed-integer *linear* reformulation using McCormick inequalities. However, as mentioned in section 3, the best global solvers for the Max-Cut problems are Biq Mac and BiqCrunch, which use specially tailored first-order methods to compute the semidefinite lower bounds. As reported in various studies [9–11], methods based on QCR techniques can solve problems up to 100 variables to global optimality, when the structure of feasible region is relatively simple, e.g., with linear constraints and binary variables. An exception is a cardinality constrained portfolio selection problem studied in [58], where the authors can handle instances up to 400 variables. However, the upper bound for the cardinality constraint is relatively small in their computational setting.

## 5 SDP techniques for cut generation

Due to the fact that it is a burden to carry the matrix variable $X$ at every node of a branch-and-bound tree, an attractive idea is to use SDP techniques to generate convex cutting constraints in the original variable space. This idea was first proposed in [49], and later also used in [25]. We again present a unified framework here, which generalizes both works. To start, we first reformulate (MIQCP) to the following form,

$$\min_{x} \; q^T x \;\; s.t. \;\; x \in \mathcal{F}, \tag{22}$$

assuming that the quadratic term in the objective is moved into constraints. Next, we assume that using some techniques mentioned in section 2, an SDP relaxation for $\mathcal{F}$ is obtained,

$$\mathcal{F} \subseteq \mathcal{Q} := \left\{ x \,\middle|\, \exists X, \begin{array}{r} \mathcal{A}(X) + Bx \leq b \\ X - xx^T \succeq 0 \end{array} \right\}.$$

Note that $\mathcal{Q}$ is a lifted convex set *projected* onto the space of original variables $x$. Optimizing over $\mathcal{Q}$ amounts to solving an SDP. To avoid this, one may construct a convex set $\hat{\mathcal{F}}$ (such that $\mathcal{Q} \subseteq \hat{\mathcal{F}}$) using only the original variable $x$ (or with small amount of lifted variables), and iteratively strengthen $\hat{\mathcal{F}}$ by adding "cutting surfaces" that are valid for $\mathcal{Q}$. Provided $\bar{x} \in \hat{\mathcal{F}}$, it is easy to see that $\bar{x} \in \mathcal{Q}$ if and only if the optimal value for the following SDP is non-positive:

$$\begin{aligned} \eta^* := \min_{\eta, X} \; & \eta \\ s.t. \; & \mathcal{A}(X) + B\bar{x} \leq b + \eta e \\ & X - \bar{x}\bar{x}^T \succeq 0. \end{aligned} \tag{$SEP_P$}$$

where $e$ is the all-one vector of proper dimension. The corresponding dual SDP is

$$\begin{aligned} \max_{y \geq 0} \; & \langle \mathcal{A}^* y, \bar{x}\bar{x}^T \rangle + (B\bar{x} - b)^T y \\ s.t. \; & \mathcal{A}^* y \succeq 0, \\ & e^T y = 1. \end{aligned} \tag{$SEP_D$}$$

Since the primal problem is guaranteed to be strictly feasible, $\bar{x} \notin \mathcal{Q}$ if and only if there exists $\bar{y}$ feasible in $(SEP_D)$ such that $\langle \mathcal{A}^* y, \bar{x}\bar{x}^T \rangle + (B\bar{x} - b)^T y > 0$. On the other hand, the inequality

$$x^T (\mathcal{A}^* \bar{y}) x + (B^T \bar{y})^T x - b^T \bar{y} \leq 0 \tag{23}$$

is convex and valid for any $x \in \mathcal{Q}$. Therefore, if $\bar{x} \notin \mathcal{Q}$ ($SEP_D$) provides a way to generate violated cuts to separate $\bar{x}$ from $\mathcal{Q}$. Further, ($SEP_D$) usually has some structure that could be potentially exploited in deriving separation heuristic algorithms.

We remark that there is some freedom in choosing the vector multiplied by $\eta$ in ($SEP_P$). In principle, one could replace the all-ones vector $e$ with any vector $a$ with nonnegative entries, as long as ($SEP_P$) remains strictly feasible. This amounts to using a normalization constraint $a^T y = 1$ in ($SEP_D$) to replace the $\ell$-1 normalization $e^T y = 1$.

As the main goal of ($SEP_D$) is to generate a valid cut, it is unnecessary to solve ($SEP_D$) to high precision. Heuristic algorithms are typically used to solve ($SEP_D$) approximately, and such heuristics may be terminated as soon as a dual feasible solution $\bar{y}$ with a sufficiently positive objective value is obtained.

Provided that an initial convex relaxation for (22) in the space of original variables (or with small number of lifted variables) is available, the aforementioned procedure can be used as a cut-generation machinery to iteratively add convex quadratic inequalities in the form of (23) and strengthen the convex relaxation. To construct an initial convex relaxation $\hat{\mathcal{F}}$, the authors in [49] split each quadratic form in the constraints into convex and concave parts using eigenvalue factorization, and derived linear relaxations for the concave part. Various other splitting strategies were studied in [29].

In [49], the authors proposed constructing the semidefinite constrained set $\mathcal{Q}$ to include the following two classes of inequalities

- Direct linearization of quadratic constraints in $\mathcal{F}$ by using $X$, i.e., if $x^T Qx + q^T x \leq d$ is one of the constraints in $\mathcal{F}$, then $\langle Q, X\rangle + q^T x \leq d$ is added into $\mathcal{Q}$;
- RLT inequalities obtained by multiplying variables bounds.

They tested this idea on a specific class of problem called BoxQP, where the feasible region $\mathcal{F}$ is described by a single nonconvex constraint and variable bounds

$$\min_{(v,x)\in\mathbb{R}^n} \left\{ v \,\middle|\, \begin{array}{r} x^T Qx + q^T x - v \leq 0, \\ x \in [0,1]^n \end{array} \right\}.$$

In their version of ($SEP_D$), they replaced $e$ with a vector that comprises of entries 1 for the constraints obtained as linearization of quadratic constraints in $\mathcal{F}$, and entries 0 for the RLT inequalities. Their separation heuristic algorithm explictly exploited the structure of RLT inequalities, and showed that all of the dual variables corresponding the RLT inequalities can be removed by replacing the objective function with a convex nonsmooth function. Then for the special case of BoxQP, the separation problem ($SEP_D$) is reduced to a one-dimensional convex minimization problem, which is then solved approximately by a specialized bisection search algorithm.

Another recent work that uses the SDP cut generation idea is [25]. The author focused on the problem with one nonconvex quadratic function and separable variable structure

$$\min_{(v,x)\in\mathbb{R}^n} \left\{ v \,\middle|\, \begin{array}{c} x^T Qx + q^T x - v \le 0, \\ x_i \in S_i,\ i = 1,\ldots,n \end{array} \right\},$$

where $S_i$ is assumed to be a union of finitely many intervals or points in $\mathbb{R}$. It is clear that BoxQP is a special case of this problem. The author constructed the semidefinite constrained set $\mathcal{Q}$ by including the linearized inequality $\langle Q, X\rangle + q^T x - v \le 0$ as well as the constraints to describe the two-dimensional convex hull condition

$$(x_i, X_{ii}) \in \mathbf{conv}\left\{(x_i, x_i^2)\middle| x_i \in S_i\right\}. \tag{24}$$

Since the only nonlinear constraint needed to describe this convex hull is $X_{ii} \ge x_i^2$, which is implied by the semidefinite constraint $X \succeq xx^T$, only finitely many linear inequalities on $(x_i, X_{ii})$ are needed to represent (24). Finally, for the normalization vector $a$, the author used a vector with entry 1 for the dual variable corresponding to $x^T Qx + q^T x - v \le 0$ and all other entries 0. Then the separation SDP ($SEP_D$) has a very special structure that is closely related to the well-known SDP relaxations for the Max-Cut problem (see [47] and the references therein). Precisely, the constraints of ($SEP_D$) becomes

$$Q + \mathbf{diag}(y) \succeq 0,$$

i.e., ($SEP_D$) seeks an optimal diagonal perturbation to convexify $Q$. The author then proposed a coordinate-minimization algorithm to solve the barrier function penalization of ($SEP_D$) while the penalization parameter is updated adaptively. Their computation experiment indicates that this approach, when applied to BoxQP instances, although theoretically weaker than the approach in [49], reduces the gap much faster in a cutting-plane framework. One of the main reasons seems to be that the heuristic algorithm does not use eigenvalue decompositions in each iteration, hence much faster than the heuristics used in [49].

Computationally, promising results of using SDP techniques as cut generation only appear on BoxQP instances [49], and nonconvex quadratic programs with separable constraints [25], and there has not been any study that incorporates these generated cuts into a branch-and-bound algorithm. We remark that the framework of ($SEP_P$) and ($SEP_D$) presented here is general, and in principle can be applied to any semidefinite constrained set $\mathcal{Q}$ used as a relaxation of $\mathcal{F}$. In many cases, the matrices involved in the constraint $\mathcal{A}^*y \succeq 0$ are sparse and/or low-rank. This type of structure, for example, can be exploited by a general-purpose interior point algorithm for SDP proposed in [8], which is implemented in the DSDP software [6, 7]. Also, the coordinate-minimization idea in [25] can be extended to solve ($SEP_D$) provided that all matrices in $\mathcal{A}^*y$ are of low-rank. Note that even if the low-rank requirement is not satisfied, high rank data matrices in the linear transformation $\mathcal{A}^*(\cdot)$ can be potentially decomposed into rank-1 matrices using eigen-reformulation and some additional variables. It would be interesting to further explore these ideas to construct better cut-generation procedures for general MIQCPs using various SDP relaxations.

## 6 Final remarks

We surveyed some recent advances of using semidefinite programming techniques to globally solve MIQCP. Although theoretical and computational evidence exist that SDP techniques can provide strong convex relaxations for various MIQCP problems, most convincingly successful adoption of SDP techniques to solve to global optimality is still restricted on structured problems in combinatorial optimization (such as those considered in [38, 47]). Extending these techniques to general MIQCP would be an interesting direction. In the second class of methods we survey, quadratic convex reformulations, information of semidefinite relaxations is only exploited at the root node, it would be interesting to answer whether updating reformulations after branching can make a positive impact to the full solution process. Finally, when quadratic forms in a MIQCP problem are sparse or "almost convex," it may not be beneficial to employ more expensive SDP techniques (one example is the Max-Cut problem of sparse graphs [47]). Therefore in an ideal global solver for MIQCP, it would be wise to incorporate SDP techniques only when necessary. The final class of methods, SDP techniques for cut generation, could be especially flexible in meeting this demand.

**Acknowledgements** We greatly appreciate two anonymous reviewers for careful reading and constructive suggestions that greatly improved this paper.

## References

1. Achterberg, T.: Scip: solving constraint integer programs. Math. Program. Comput. **1**(1), 1–41 (2009). http://mpc.zib.de/index.php/MPC/article/view/4
2. Alizadeh, F.: Interior point methods in semidefinite programming with applications to combinatorial optimization. SIAM J. Optim. **5**, 13–51 (1995)
3. Anstreicher, K.M.: Semidefinite programming versus the reformulation-linearization technique for nonconvex quadratically constrained quadratic programming. J. Glob. Optim. **43**, 471–484 (2009)
4. Bao, X., Sahinidis, N.V., Tawarmalani, M.: Semidefinite relaxations for quadratically constrained quadratic programming: A review and comparisons. Math. Program. B **129**, 129–157 (2011)
5. Belotti, P.: COUENNE: a user's manual. Technical Report, Department of Mathematical Sciences, Clemson University (2012)
6. Benson, S.J., Ye, Y.: DSDP5: software for semidefinite programming. Technical Report ANL/MCS-P1289-0905, Mathematics and Computer Science Division, Argonne National Laboratory, Argonne, IL (2005). http://www.mcs.anl.gov/~benson/dsdp. Submitted to ACM Transactions on Mathematical Software
7. Benson, S.J., Ye, Y.: DSDP5 user guide — software for semidefinite programming. Technical Report ANL/MCS-TM-277, Mathematics and Computer Science Division, Argonne National Laboratory, Argonne (2005). http://www.mcs.anl.gov/~benson/dsdp
8. Benson, S.J., Ye, Y., Zhang, X.: Solving large-scale sparse semidefinite programs for combinatorial optimization. SIAM J. Optim. **10**(2), 443–461 (2000)

9. Billionnet, A., Elloumi, S.: Using a mixed integer quadratic programming solver for the unconstrained quadratic 0-1 problem. Math. Program. A **109**, 55–68 (2007)
10. Billionnet, A., Elloumi, S., Plateau, M.C.: Improving the performance of standard solvers for quadratic 0-1 programs by a tight convex reformulation: the QCR method. Discret. Appl. Math. **157**, 1185–1197 (2009)
11. Billionnet, A., Elloumi, S., Lambert, A.: Extending the QCR method to general mixed-integer programs. Math. Program. A **131**, 381–401 (2012)
12. Bomze, I.M., de Klerk, E.: Solving standard quadratic optimization problems via linear, semidefinite and copositive programming. J. Global Optim. **24**(2), 163–185 (2002). Dedicated to Professor Naum Z. Shor on his 65th birthday
13. Bonami, P., Biegler, L.T., Conn, A.R., Cornuéjols, G., Grossmann, I.E., Laird, C.D., Lee, J., Lodi, A., Margot, F., Sawaya, N., Wächter, A.: An algorithmic framework for convex mixed integer nonlinear programs. Discre. Optim. **5**, 186–204 (2008)
14. Burer, S.: Copositive programming. In: Anjos, M., Lasserre, J.B. (eds.) Handbook of Semidefinite, Cone and Polynomial Optimization: Theory, Algorithms, Software and Applications, Chap. 5. Springer, Berlin (2010)
15. Burer, S.: Optimizing a polyhedral-semidefinite relaxation of completely positive programs. Math. Prog. Comput. **2**, 1–19 (2010)
16. Burer, S., Dong, H.: Separation and relaxation for cones of quadratic forms. Math. Prog. A. **137**(1), 343–370 (2013)
17. Burer, S., Letchford, A.N.: On nonconvex quadratic programming with box constriants. SIAM J. Optim. **20**(2), 1073–1089 (2009)
18. Burer, S., Monteiro, R.: A nonlinear programming algorithm for solving semidefinite programs via low-rank factorization. Math. Program. B **95**, 329–357 (2003)
19. Burer, S., Saxena, A.: The MILP road to MIQCP. In: Mixed Integer Nonlinear Programming: The IMA Volumes in Mathematics and its Applications, vol. 154, pp. 373–405. Springer, Berlin (2012)
20. Burer, S., Vandenbussche, D.: Globally solving box-constrained nonconvex quadratic programs with semidefinite-based finite branch-and-bound. Comput. Optim. Appl. **43**(2), 181–195 (2009)
21. Burer, S.A.: On the copositive representation of binary and continuous nonconvex quadratic programs. Math. Program. **120**, 479–495 (2009)
22. Burer, S.A., Anstreicher, K.M.: Second-order-cone constraints for extended trust-region subproblems. SIAM J. Optim. **23**, 432–451 (2013)
23. Burer, S.A., Chen, J.: Globally solving nonconvex quadratic programming problems via completely positive programming. Math. Program. Comput. **4**, 33–52 (2012)
24. Dong, H.: Symmetric tensor approximation hierarchies for the completely positive cone. SIAM J. Optim. **23**(3), 1850–1866 (2013)
25. Dong, H.: Relaxing nonconvex quadratic functions by multiple adaptive diagonal perturbations. Optimization Online http://www.optimization-online.org/DB_HTML/2014/03/4274.html (2014)
26. Dong, H., Anstreicher, K.: Separating Doubly Nonnegative and Completely Positive Matrices. Math. Program. A. **137**(1), 131–153 (2013)
27. Drewes, S., Ulbrich, S.: Subgradient based outer approximation for mixed integer second order cone programming. In: Lee, J., Leyffer, S. (eds.) Mixed Integer Nonlinear Programming. The IMA Volumes in Mathematics and its Applications, vol. 154, pp. 41–59. Springer, Berlin (2012)
28. Duran, M.A., Grossmann, I.: An outer-approximation algorithm for a class of mixed-integer nonlinear programs. Math. Program. **36**, 307–339 (1986)
29. Fampa, M., Lee, J., Melo, W.: On global optimization with indefinite quadratics. Isaac Newton Institute Preprint NI13066 (2013)
30. Faye, A., Roupin, F.: Partial Lagrangian relaxation for general quadratic programming. 4OR **5**, 75–88 (2007)

31. Galli, L., Letchford, A.N.: A compact variant of the QCR method for quadratically constrained quadratic 0-1 programs. Optim. Lett. **8**, 1213–1224 (2014)
32. Geomans, M.X., Williamson, D.P.: Improved approximation algorithm for maximum cut and satisfiability problems using semidefinite programming. J. ACM **42**, 1115–1145 (1995)
33. Gupte, A., Ahmed, S., Dey, S., Cheon, M.: Pooling problem. In: Furman, K., Song, J. (eds.) Optimization and Analytics in the Oil and Gas Industry, International Series in Operations Research and Management Science. Springer, Berlin (2015)
34. Gurobi Optimization Inc., Houston, T.: Gurobi optimizer reference manual version 3.0. (2010)
35. Hammer, P.L., Rubin, A.A.: Some remarks on quadratic programming with 0-1 variables. RAIRO - Oper. Res. **4**(3), 67–79 (1970)
36. Helmberg, C., Rendl, F.: A spectral bundle method for semidefinite programming. SIAM J. Optim. **10**, 673–696 (2000)
37. de Klerk, E., Pasechnik, D.V.: Approximation of the stability number of a graph via copositive programming. SIAM J. Optim. **12**(4), 875–892 (2002)
38. Krislock, N., Malick, J., Roupin, F.: Improved semidefinite branch-and-bound algorithm for k-cluster. Comput. Oper. Res. (2013, Revision submitted). https://hal.archives-ouvertes.fr/hal-00717212
39. Krislock, N., Malick, J., Roupin, F.: Improved semidefinite bounding procedure for solving Max-Cut problems to optimality. Math. Program. **143**(1-2), 61–86 (2014)
40. Letchford, A.N., Sørensen, M.M.: A new separation algorithm for the boolean quadric and cut polytopes. Discret. Optim. **14**, 61–71 (2014)
41. Lovász, L.: On the shannon capacity of a graph. IEEE Trans. Inf. Theory **25**, 1–7 (1979)
42. Malick, J.: The spherical constraint in boolean quadratic programs. J. Glob. Optim. **39**, 609–622 (2007)
43. Malick, J., Roupin, F.: On the bridge between combinatorial optimization and nonlinear optimization: a family of semidefinite bounds for 0–1 quadratic problems leading to quasi-Newton methods. Math. Program. B. **140**(1), 99–124 (2013)
44. Misener, R., Floudas, C.A.: Global optimization of mixed-integer quadratically constrained quadratic programs (MIQCQP) through piecewise-linear and edge-concave relaxations. Math. Program. B **136**, 155–182 (2012)
45. Misener, R., Floudas, C.A.: GloMIQO: Global mixed-integer quadratic optimizer. J. Glob. Optim. **57**, 3–30 (2013)
46. Parrilo, P.A.: Structured semidefinite programs and semialgebraic geometry methods in robustness and optimization. Ph.D. Thesis, California Institute of Technology (2000)
47. Rendl, F., Rinaldi, G., Wiegele, A.: Solving Max-Cut to optimality by intersecting semidefinite and polyhedral relaxations. Math. Program. **121**(2), 307 (2010)
48. Sahinidis, N.V.: BARON: a general purpose global optimization software package. J. Glob. Optim. **8**, 201–205 (1996)
49. Saxena, A., Bonami, P., Lee, J.: Convex relaxations of non-convex mixed integer quadratically constrained programs: projected formulations. Math. Program. A **130**(2), 359–413 (2010)
50. Shor, N.: Dual quadratic estimates in polynomial and boolean programming. Ann. Oper. Res. **25**, 163–168 (1990)
51. Sotirov, R.: SDP relaxations for some combinatorial optimization problems. In: Anjos, M.F., Lasserre, J.B. (eds.) Handbook of Semidefinite, Cone and Polynomial Optimization: Theory, Algorithms, Software and Applications. International Series in Operations Research and Management Science, vol. 166, chap. 27, pp. 795–819. Springer, Berlin (2012)
52. Sponsel, J., Dür, M.: Factorization and cutting planes for completely positive matrices by copositive projection. Math. Prog. **143**(1-2), 211–229 (2012)
53. Sturm, J.F., Zhang, S.: On Cones of Nonnegative Quadratic Functions. Math. Oper. Res. **28**(2), 246–267 (2003)
54. Todd, M.J.: Semidefinite optimization. Acta Numer. **10**, 515–560 (2001)
55. Vandenberghe, L., Boyd, S.: Semidefinite programming. SIAM Rev. **38**, 49–95 (1996)

56. Vielma, J.P., Ahmed, S., Nemhauser, G.L.: A lifted linear programming branch-and-bound algorithm for mixed integer conic quadratic programs. INFORMS J. Comput. **20**, 438–450 (2008)
57. Wen, Z., Goldfarb, D., Yin, W.: Alternating direction augmented Lagrangian methods for semidefinite programming. Math. Program. Comput. **2**, 203–230 (2010)
58. Zheng, X., Sun, X., Li, D.: Improving the performance of MIQP solvers for quadratic programs with cardinality and minimum threshold constraints: A semidefinite program approach. INFORMS J. Comput. **26**(4), 690–703 (2014)

# A few strong knapsack facets

**Sunil Chopra, Sangho Shim, and Daniel E. Steffy**

**Abstract** We perform a shooting experiment for the knapsack facets and observe that $1/k$-facets are strong for small $k$; in particular, $k$ dividing 6 or 8. We also observe spikes of the size of $1/k$-facets when $k = n$ or when $k + 1$ divides $n + 1$. We discuss the strength of the $1/n$-facets introduced by Aráoz et al. (Math Program 96:377–408, 2003) and the knapsack facets given by Gomory's homomorphic lifting.

A general integer knapsack problem is a knapsack subproblem where a portion, often a significant majority, of the variables are missing from the master knapsack problem. The number of projections of $1/k$-facets on a knapsack subproblem of $l$ variables is $O(l^{\lceil k/2 \rceil})$, note that this is independent of the size of the master problem. Since $1/k$-facets are strong for small $k$, we define the $1/k$-inequalities which include the $1/d$-facets with $d$ dividing $k$ and fix $k$ to be a small constant such as $k = 6$ or $k = 8$. We develop an efficient way of enumerating violated valid $1/k$-inequalities. For each violated $1/k$-inequality, we determine its validity by solving a small integer programming problem, the size of which depends only on $k$.

**Keywords** Knapsack problem • Master knapsack polytope • Facets • Shooting experiment • Cutting planes

**Mathematics subject classification (2010):** 90C10 Integer Programming

---

This chapter was prepared for publication in MOPTA 2014 Proceedings.

S. Chopra
Kellogg Graduate School of Management, Northwestern University, Leverone Hall, Evanston, IL 60208, USA

S. Shim (✉)
Department of Engineering, Robert Morris University, Moon Twp, PA 15108, USA
e-mail: shim@rmu.edu

D.E. Steffy
Department of Mathematics and Statistics, Oakland University, Rochester, MI 48309, USA

 
B. Defourny, T. Terlaky (eds.), *Modeling and Optimization: Theory and Applications*, Springer Proceedings in Mathematics & Statistics 147,
DOI 10.1007/978-3-319-23699-5_4

# 1 Introduction

The *master knapsack problem* of order $n$ is defined to be

$$\max \ vt \tag{1}$$

$$st \ \sum_{i=1}^{n} it_i = n \tag{2}$$

$$t \geq 0 \tag{3}$$

$$t_i \text{ are integers,} \tag{4}$$

where $v \geq 0$ is a row vector of length $n$, and $t$ is a column vector of $n$ variables. Observe that equation (2) contains all integer coefficients from 1 to $n$. The problem given by (1)–(4) is known as the *master knapsack problem*, $K(n)$.

The convex hull of the solutions to $K(n)$ is denoted by $P(K(n))$ and referred to as the *master knapsack polytope*. The dimension of $P(K(n))$ is $n-1$ and the non-negativity constraints (3) are facet-defining for $i \geq 2$ (see Shim [16] and Shim and Johnson [17]). We call the facet-defining non-negativity constraints *trivial* facets. The other facets are called *knapsack facets*. Since $P(K(n))$ is not full dimensional, each knapsack facet has infinitely many representations. Throughout this paper, we consider the representation $\xi t \leq 1$ with $\xi_1 = 0$ and $\xi_n = 1$, where $\xi$ is the length $n$ row vector of coefficients. Overloading our notation we use $\xi$ as a representative of $\xi t \leq 1$ and refer to this vector itself as a knapsack facet. A knapsack facet $\xi t \leq 1$ is called a $1/k$-*facet* if $k$ is the smallest possible integer such that

$$\xi_i \in \{0/k, 1/k, 2/k, \ldots, k/k\} \cup \{1/2\}. \tag{5}$$

Note that $k$ is the least common multiple of the denominators of the irreducible fractions $\xi_i$ except $\xi_i = 1/2$. The set of indices $i$ where the coefficients $\xi_i = 1/2$ is referred to as the *half landing*. The half-landing of a knapsack facet is known in [18] to have indices $i$ with $n/3 < i < 2n/3$. We call a $1/k$-facet *strict* if $m/k$ appears as a coefficient for every $m \in \{1, 2, \ldots, k\} \setminus \{k/2\}$.

The remainder of the paper is organized as follows. In Section 2, we introduce some specific classes of knapsack facets to be further discussed and studied. In Section 3, we describe the results of a shooting experiment and observe that $1/k$-facets are important for $k$ dividing 6 or 8. We also observe importance of $1/k$-facets when $k = n$ or when $k+1$ divides $n+1$.

Knapsack subproblems are restrictions of the master knapsack problem where the variables $t_i$ only appear for $i \in L$ where $L \subseteq \{1, \ldots, n\}$, i.e., some of the coefficients do not appear in the equation. The convex hull of the feasible solutions to a knapsack subproblem is given by

$$P(K(n)) \cap \{t : t_i = 0 \ \ \forall i \notin L\}.$$

Knapsack subproblems provide a connection between the master knapsack problem and many single row relaxations of general integer programs. The projections of the knapsack facets are valid inequalities for a knapsack subproblem.

In Section 4, we develop an algorithms to separate and enumerate $1/k$-facets for some small values of $k$ for knapsack subproblems. The number of projections of $1/k$-facets for a knapsack subproblem of $l$ variables is $O(l^{\lceil k/2 \rceil})$, which is independent of the size of the master problem. We therefore are focused on the use of separation and enumeration algorithms for the knapsack subproblems whose running time are independent of $n$.

In Section 5, we discuss a notion of the strength of a knapsack facet and analyze the strength of a class of $1/n$-facets, which were observed to be important in the shooting experiment. In Section 6, we discuss some special classes of knapsack subproblems where $1/k$-facets with small $k$ are likely to be ineffective and suggest alternative solution methods, such as group relaxations. Finally, Section 7 gives concluding remarks.

Following Gomory, several authors have studied polyhedra corresponding to binary and cyclic groups chracterizing the facets by subadditive relations in Fulkerson's framework of blocking polyedra. For example, Gomory and Johnson [9–11] defined two-slope facets for master cyclic group polyhedra, and Cornuejols and Molinaro [4] and Basu, Hildebrand, Koppe and Molinaro [3] have defined other families of facets for such polyhedra, including three-slope and $(k+1)$-slope facets. Shu, Chopra, Johnson and Shim [19] gave a new class of 1/3- and 1/4-facets with no 0-valued coefficient for master binary group polyhedra. Araoz, Evans, Gomory and Johnson [1] studied the relation between cyclic group and knapsack facets, and defined strong families of knapsack facets that come from 2-slope facets. In this paper, we analyze knapsack facets characterized by superadditive relations and show that facets with simple shapes are stronger than those with complex shapes. We develop a method of separating the simple facets and generalize the method to integer programming.

## 2 Characterization of $1/k$-facets

Aráoz [1] characterized the knapsack facets as the extreme rays of a polynomially sized system of super-additive relations. Hunsaker [14] described the knapsack facets by the extreme points of the system with fixing $\xi_1 = 0$.

**Theorem 2.1 (Aráoz [1], Hunsaker [14]).** *The coefficient vectors $\xi$ of the knapsack facets $\xi t \leq 1$ of $K(n)$ with $\xi_1 = 0$ and $\xi_n = 1$ are the extreme points of the system of linear constraints*

$$\xi_1 = 0, \tag{6}$$

$$\xi_n = 1, \tag{7}$$

$$\xi_i + \xi_{n-i} = 1 \quad \text{for } 1 \leq i \leq n/2, \tag{8}$$

$$\xi_i + \xi_j \leq \xi_{i+j} \quad \text{whenever } i + j < n. \tag{9}$$

*The feasible solutions to the system give valid inequalities $\xi t \leq 1$ for $P(K(n))$.*

We call (8) and (9) the *complementarities* and the *superadditivities*, respectively. Therefore, a knapsack facet $\xi$ is a non-decreasing sequence because

$$\xi_i = 0 + \xi_i = \xi_1 + \xi_i \leq \xi_{i+1}.$$

Although $P(K(n))$ has exponentially many facets, certain polynomially sized subsets of these facets appear to be of special importance. In the following we review some of these classes.

## 2.1 $1/k$-inequalities

In this paper, a sequence $\xi = (\xi_i)_{i=1}^n$ is called *symmetric* if the complementarities (8) hold. We call $\xi t \leq 1$ a $1/k$-*inequality* if $\xi$ is a non-decreasing symmetric sequence that satisfies (5). In general, a $1/k$-inequality need not be a valid inequality for $P(K(n))$. We see that a $1/d$-inequality is a $1/k$-inequality if $d$ is a divisor of $k$. A $1/k$-inequality $\xi$ is uniquely determined by a non-decreasing sequence $(a_m)$ where $a_m$ represents the first index $i$ with $\xi_i \geq m/k$ for $m \in \{0, 1, \ldots, k\} \cup \{k/2\}$. Observe that $k/2$ is not an integer for $k$ odd but is required to obtain the coefficient $1/2$. If $a_m = a_{m+1}$ or if $a_m = a_{m+1/2}$ with $k$ odd and $m \in \{(k-1)/2, k/2\}$, no coefficient $\xi_i$ has value $m/k$. Such a sequence $\xi$ will be denoted by $\xi^{k\text{-}(a_m)}$. Also, because of symmetry, the number of $\xi_i$'s of value $m/k$ must equal the number of those of value $(k-m)/k$. Thus, $a_m$ for $m = 1, \ldots, k/2$ are sufficient to uniquely define $\xi^{k\text{-}(a_m)}$. A $1/k$-inequality is called a $1/k$-*facet* if it is a knapsack facet and if $k$ is the smallest possible integer that satisfies (5). We remark that every facet is a $1/k$-facet for some value of $k$. However, fixing $k$ allows us to consider specific, polynomially sized, classes of facets; we are particularly interested in certain small fixed values of $k$, the importance of which has been demonstrated by Shim, Cao, and Chopra [18].

## 2.2 $1/n$-facets

Aráoz et al. [2] defined two families of knapsack facets which are equivalent to $1/n$-facets. One family is defined in Theorem 6.5 of [2] and equivalent to the $1/n$-facets $\xi^{n\text{-}(a_m)}$ given by $a_1 = \ldots = a_q = q$ and by $a_i = \lceil i \rceil$ for $q \leq i \leq n-q$ where $1 < q \leq \frac{n}{4}$.

The other family is defined in Theorem 6.3 of [2] and equivalent to the $1/n$-facets $\xi^{n\text{-}(a_m)}$ given by $a_i = i$ for $i < q$ even and $a_i = i + 1$ for $i < q$ odd and by $a_i = i$ for $q \leq i \leq n-q$ where $q \leq \frac{n}{2}$ if $n$ is even and $q \leq \frac{n-2}{3}$ if $n$ is odd. Any knapsack facet $\xi$ with $\xi_2 = \frac{2}{n}$ belongs to the family, as shown in Aráoz et al. [2].

In Section 5.3, we analyze the strength of these two families of $1/n$-facets and see that the first family appears to be important, and the second one appears to be less important, this agrees with the findings in the shooting experiment described in Section 3.

## 2.3 Group homomorphically lifted facets

A knapsack problem may be relaxed to the cyclic group problem, defined by Gomory [8]. The *cyclic group problem* $(C_n, b)$ with respect to a cyclic group $C_n$ of order $n$ and a non-zero element $b$ has a feasible region given by vectors $t$ that satisfy

$$\sum_{g \in C_n \setminus \{0\}} g t_g = b,$$

where $t_g$, $g \in C_n \setminus \{0\}$, are non-negative integer variables. The convex hull of the integer solutions $t$ of the problem was shown by Gomory [8] to be a polyhedron and is referred to as the *cyclic group polyhedron*, denoted $P(C_n, b)$. The non-negativity constraints $t_g \geq 0$, $g \neq 0$, are facets of $P(C_n, b)$ and called *trivial* facets. The non-trivial facets are denoted by $\pi t \geq \pi_b > 0$ and called *cyclic group facets*.

A *homomorphism* $\phi$ of a group $G$ into a group $H$ is a map $\phi : G \to H$ which preserves the addition; i.e.,

$$\phi(g_1 + g_2) = \phi(g_1) + \phi(g_2) \text{ for all } g_1, g_2 \in G.$$

The *kernel* $Ker(\phi)$ of the map $\phi : G \to H$ is defined to be the set of elements $g$ in $G$ which are mapped to $\phi(g) = 0$.

Gomory [8] showed a facet of a group polyhedron with a 0 coefficient can be constructed by repeating a facet of a lower dimensional group polyhedron. His lifting theorem enables us to assemble facets as building blocks for higher dimensional facets:

**Theorem 2.2.** *Let $G$ be an abelian group and let $K$ be a subgroup of $G$. Let $b \in G \setminus K$ and let $\phi : G \to G/K$ be the canonical homomorphism of $G$ onto the factor group $G/K$. If $\hat{\pi}$ is a group facet for $G/K$ with right-hand side $\hat{b} = \phi(b)$, then a group facet for $G$ with right-hand side $b$ is given by*

$$\pi(g) = \hat{\pi}(\phi(g)).$$

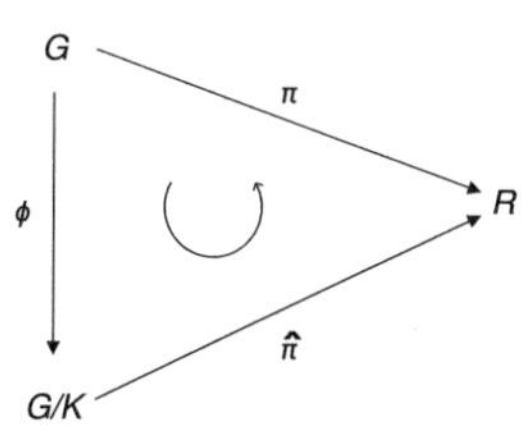


Lifted facets are shown to be important for the cyclic group polyhedron in the shooting experiment performed by Gomory, Johnson, and Evans [12].

We refer to the vector $(1/n, 2/n, \ldots, n/n)$ as *lineality* and denote it by $lin(n)$, note that this is simply the coefficient vector of the knapsack equation divided by $n$. There is a connection between the knapsack facets and the facets of the cyclic group polyhedron; namely, the knapsack facets for $P(K_n)$ are precisely the cyclic group facets which are adjacent to the lineality $lin(n)$ in $P(C_{n+1}, n)$ [2]. The following theorem states a tilting by which our representation $\xi t \leq 1$ of a knapsack facet can be converted into a cyclic group facet $\pi^{\xi} t \geq 1$ of $P(C_{n+1}, n)$.

**Theorem 2.3.** *Let* $sep(\xi) = \max_{i+j>n+1} \xi_i + \xi_j - \xi_{i+j-n-1}$ *and let*

$$\pi^{\xi} = \frac{n+1}{n \cdot sep(\xi) - (n+1)} \cdot (lin(n) - \xi) + lin(n).$$

*Then, the knapsack facet* $\xi t \leq 1$ *with* $\xi_1 = 0$ *and* $\xi_n = 1$ *can be alternatively represented by* $\pi^{\xi} t \geq 1$ *which is a cyclic group facet for* $P(C_{n+1}, n)$.

Since lifted cyclic group facets are important, we suspect the knapsack facets equivalent to lifted cyclic group facets to also be important. If $k+1$ divides $n+1$, there is a $1/k$-facet such that its equivalent cyclic group facet has a zero coefficient $\pi_i = 0$ and is a lifted facet; in particular, the repetition $\pi^{\xi}$ of $lin\left(\frac{n+1}{k+1}\right)$ is a cyclic group facet of $(C_{n+1}, n)$ having a zero coefficient. In Section 3, we observe that $1/k$-facets are strong when $k+1$ divides $n+1$.

## 3 Shooting experiment

Intuitively, the shooting experiment shoots an arrow from the origin toward the knapsack facets in a random direction sampled from the spherically uniform distribution in the non-negative orthant, and sees which facet is hit first. A random vector $v$ follows the spherically uniform distribution in the non-negative orthant if $v_i$ are the absolute values of the independent and identically normally distributed random variables with mean 0. While this conceptual process might seem to require checking exponentially many knapsack facets against the random direction, the facet hit by each shot can be determined in polynomial time by solving a single linear program. As done by Hunsaker [14], we may perform a shooting experiment by solving the *shooting linear programming problem* to maximize a random direction $v \geq 0$ over the constraints (6)–(9) with variables $\xi$. Its optimal solution $\xi$ is the coefficient vector of the knapsack facet $\xi t \leq 1$ hit by shooting in $v$.

Shooting experiments were first used in the context of the TSP by Kuhn [15], and then by Gomory, Johnson and Evans [12], Evans [6], and Dash and Günlük [5] for the master cyclic group polyhedron.

## 3.1 Minimal characterization of the knapsack facets

We identify a minimal representation of the system (6)–(9) transforming the minimal representation in Shim [16].

**Theorem 3.1.** *A minimal representation of the system (6)–(9) is the equalities in (6)–(8) and the inequalities (9) replaced by*

$$\xi_i + \xi_j \leq \xi_{i+j} \textit{ for } i \leq j < i+j < n/2, \tag{10}$$

$$\xi_i + \xi_j + \xi_{n-i-j} \leq 1 \textit{ for } i \leq j \leq n-i-j < n/2, \tag{11}$$

$$\textit{and } 2\xi\left(\frac{n}{4}\right) \leq \xi\left(\frac{n}{2}\right) = \frac{1}{2} \textit{ if } n \equiv 0 \mod 4. \tag{12}$$

The shooting experiment is also equivalent to solving the linear programming problem to maximize

$$\sum_{1<i<n/2} (v_i - v_{n-i})\xi_i$$

over the system (10)–(12). Its optimal solution ($\xi_i : 1 < i < n/2$) can be lifted by complementarities (8) to the coefficient vector $\xi = (\xi_i : i = 1, \ldots, n)$ of a knapsack facet $\xi t \leq 1$. The number of variables of the reduced shooting LP is half that of the original shooting LP and the number of constraints reduces to one third. Hence, Theorem 3.1 allows performing the shooting experiment for a larger $n$.

## 3.2 Concentration on $1/k$-facets with k dividing 6 or 8

Shim and Johnson [17] performed a shooting experiment firing off 10,000 shots for small order $n \leq 20$ and identified a pattern of the most hit knapsack facets. In every shooting experiment, the most hit facet was always the 1-facet of rank 0 and all the 1-facets absorbed more than 50% of hits except $n = 6, 12, 18$. This numerical experiment suggested that if we look at the knapsack facets from the origin, the 1-facets, a family of linear size in $n$, will dominate our field of view, despite the fact that exponentially many facets are needed to describe $P(K(n))$. The 1-facets are shown in [17] to be adjacent to each other.

Figure 1 is a result of our new shooting experiment firing off one million shots for large order $n \leq 150$ and confirms that the 1-facets absorb more than 50% of the hits except for the cases $n = 6, 12, 18, 24$. The horizontal axis of the figure indicates $n = 5, \ldots, 150$ and the vertical axis indicates the number of hits absorbed by the 1-facets out of a million shots for each $n$. In the figure we see that the $1/k$-facets with $k$ dividing 6 or 8 absorb more than 75% of shots.

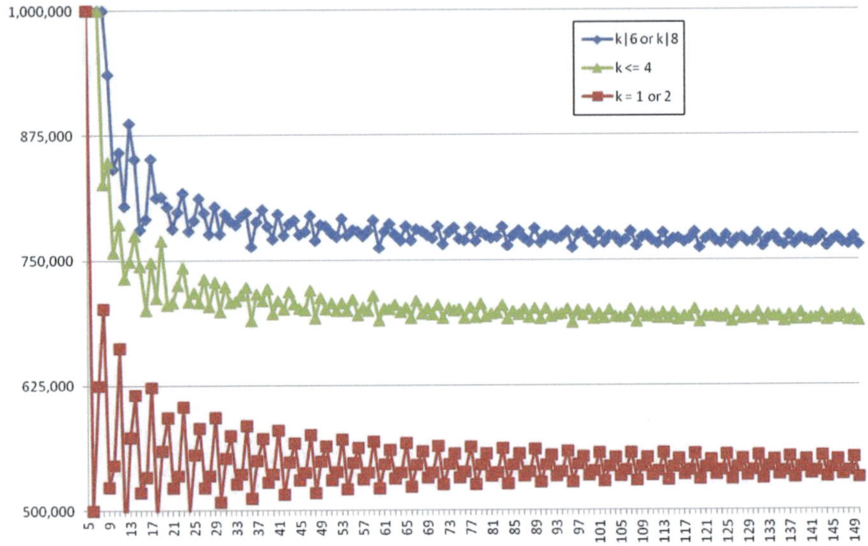


**Fig. 1** The number of hits absorbed by $1/k$-facets of $K(n)$ out of one million shots for each $n \leq 150$

### *3.3 Decrease and oscillation of the size of $\mathbf{1/k}$-facets*

Figure 2 depicts the result of the shooting experiment for $K(29)$, the graph plots the number of shots absorbed by $1/k$-facets and lifted $1/k$-facets for each value of $k$ up to 30. Here we see that the number of $1/k$-facets hit decreases as $k$ grows and oscillates; i.e., $1/k$-facets are hit more often with $k$ even than those with $k$ odd. The worst case analysis of [18] suggests that the size of regular $1/k$-facets decreases and oscillates in the same manner. The result of shooting experiment goes together with the worst case analysis and we provide theoretical support for the results of the shooting experiment.

### *3.4 Spikes at $\mathbf{1/n}$-facets and lifted facets*

In Figure 2, we see two kinds of spikes; one at $k = n$ and the other at $k$ with $k+1$ dividing $n+1$. The $1/n$-facets are frequently hit when $n$ is small, and somewhat less hit as $n$ grows. We are more interested in $1/k$-facets with small $k$. However, lifted facets are frequently hit for all values of $n$; in particular $1/k$-facets that correspond to repetitions of $lin(d)$ with $d = (n+1)/(k+1)$ when tilted to the $\pi$ notation of $(C_{n+1}, n)$ are frequently hit.

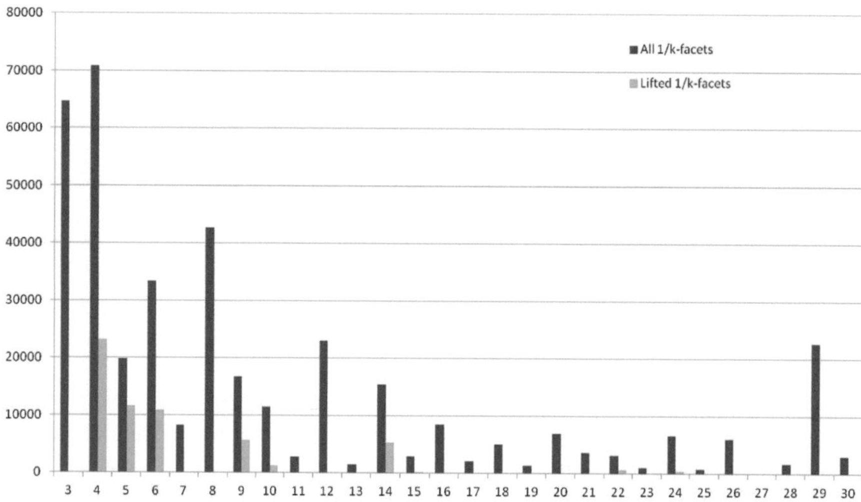


**Fig. 2** The number of hits absorbed by $1/k$-facets, and lifted $1/k$-facets of $K(29)$ out of one million shots. (Case $k = 2$ excluded from plot)

# 4 Separation and enumeration for subproblems

Recall that a knapsack subproblem is a restriction of the master knapsack problem that is missing some terms in the knapsack equation (2); *i.e.*,

$$\sum_{i \in L} it_i = n, \tag{13}$$

where $L$ is a non-empty subset of $\{1, \ldots, n\}$. We are especially interested in cases when the dimension $l = |L|$ of the subproblem is much smaller than the size $n$ of the master problem. The knapsack facets are valid inequalities for a knapsack subproblem.

## *4.1 Separation of the projections of 1-facets*

Although the master knapsack problem has $\Theta(n)$ 1-facets, a knapsack subproblem has only $O(l)$ unique projected 1-facets. We now develop a separation algorithm that will identify the 1-facet most violated by a given nonnegative solution $\hat{t}$ in time linear in $l$. Recall that a 1-facet $\xi^{2\text{-}(a_1,a_2)}$ is defined uniquely by the number $a_1$ which is known to satisfy $n/3 < a_1 \leq (n+1)/2$.

If the following quantity is greater than zero, it gives the amount by which a nonnegative solution $\hat{t}$ violates the 1-facet $\xi^{2\text{-}(a_1,a_2)}$ (if non-positive, it is the slack by which $\hat{t}$ satisfies the 1-facet):

$$VIOL(a_1) = \xi^{2\text{-}(a_1,a_2)}\hat{t} - 1 = \left( \sum_{i\in L, a_1 \le i < n-a_1} \hat{t}_i/2 + \sum_{i \in L, i \ge n-a_1} \hat{t}_i \right) - 1. \tag{14}$$

To find the most violated 1-facet we must determine

$$\max_{n/3 < a_1 \le (n+1)/2} VIOL(a_1). \tag{15}$$

However, in order to find $a_1$ maximizing the quantity in formula (15) we observe that many values of $a_1$ need not be considered because $L$ does not contain all indices between 1 and $n$. Namely, it is sufficient to consider $a_1 \in X$ where $X = \{x : n/3 < x \le (n+1)/2$ and either $x \in L$ or $n - x \in L\}$. Essentially, $X$ gives a set of indices that are feasible for $a_1$ and also correspond to indices where $L$ has a nonzero component in position $a_1$, or the complementary position $a_2 = n - a_1$. (There is one exceptional case when $X$ is empty, in such case it is enough to consider the 1-facet given by $a_1 = \lfloor (n+1)/2 \rfloor$, which can be easily checked.)

We assume that the values in $X = \{x_1, \ldots, x_m\}$ are in sorted order. Note that $VIOL(x_1)$ may be computed in $O(l)$ time using formula (14). Additionally, if $1 < i \le m$ then

$$VIOL(x_i) = VIOL(x_{i-1}) + \frac{1}{2}(\hat{t}_{x_i} - \hat{t}_{n-x_i}) \tag{16}$$

where $\hat{t}_k$ is understood to be zero whenever $k \notin L$. Noting that $|X| = O(l)$ we see that this naturally leads to a dynamic programming style $O(l)$ time algorithm to evaluate formula (15) and find the 1-facet most violated by $\hat{t}$. A simple modification to this algorithm would also allow us to enumerate all violated inequalities from this class.

## 4.2 *Enumeration of the violated projections of* $1/k$*-inequalities*

The number of the projections of the $1/k$-inequalities is $O\left(l^{\lceil k/2 \rceil}\right)$ which is independent of $n$, the size of the master knapsack problem $K(n)$. If we enumerate the $1/k$-inequalities one by one and compute $LHS = \xi_L t_L$ in linear time $O(l)$ to see violation or $LHS > 1$ for each $1/k$-inequality, the total time for enumerating the violated $1/k$-inequalities is $O\left(l^{\lceil k/2 \rceil + 1}\right)$. In a similar manner to Section 4.1, we can enumerate the violated projections of $1/k$-inequalities keeping time $O\left(l^{\lceil k/2 \rceil}\right)$ by updating $\xi_L t_L$ from iteration to iteration (in constant time) instead of computing it from scratch for each $\xi_L$. However, as noted previously not all $1/k$-inequalities are valid; in the next section, we describe a very small integer program that can be used to determine if $\xi_L$ defines a valid inequality.

## 4.3 Validity of the projections of $1/k$-inequalities

For each violated projection $\xi_L$ of a $1/k$-inequality, we must check if there is a coefficient vector $\xi$ of a valid inequality satisfying (6)–(9), implying the validity of $\xi_L$. The following gives a polyhedral description of the necessary relationships between the elements of the sequence $(a_m)$, defining the coefficients of a $1/k$-inequality $\xi^{k\text{-}(a_m)}$.

**Lemma 4.1.** *A $1/k$-inequality $\xi^{k\text{-}(a_m)}$ satisfies (6)–(9) if and only if*

$$2 \le a_{m_1} \le a_{m_2} \le (n+1)/2 \quad \textit{for } m_1 \le m_2, \tag{17}$$

$$a_m + a_{k+1-\lceil m \rceil} = n+1 \quad \textit{for } m \le k/2, \tag{18}$$

$$a_{m_1} + a_{m_2} \ge a_{\lceil m_1+m_2 \rceil} \quad \textit{for all } m_1 \le m_2 \textit{ with } \lceil m_1+m_2 \rceil \le k. \tag{19}$$

In order to check if there is $\xi$ satisfying (6)–(9), we can check if there is an integer solution $(a_m)$ satisfying (17)–(19). Note that the system has $\lceil k/2 \rceil$ variables and $O(k^2)$ constraints. The number of variables and the number of constraints are independent of $l$ and constant if $k$ is fixed to be a constant. We now give two examples illustrating how this technique can be applied.

*Example 4.2.* A $1/6$-inequality $\xi^{6\text{-}(a_m)}$ is valid if there is an integer solution $(a_m)$ satisfying

1. Complementarities: $a_1 + a_6 = n+1$, $a_2 + a_5 = n+1$, $a_3 + a_4 = n+1$
2. Non-decreasing: $2 \le a_1 \le a_2 \le a_3 \le (n+1)/2$
3. Sub-additivities: $2a_1 \ge a_2$, $a_1 + a_2 \ge a_3$; $a_1 + 2a_3 \ge n+1$, $2a_2 + a_3 \ge n+1$
4. Defining: $x_{i_1-1} + 1 \le a_1 \le x_{i_1}$, $x_{i_2-1} + 1 \le a_2 \le x_{i_2}$, $x_{i_3-1} + 1 \le a_3 \le x_{i_3}$

where $x_{i_1}, x_{i_2}, x_{i_3} \in X$ are the smallest indices $x \in X$ of $\xi_x \ge 1/6, 2/6, 3/6$.

*Example 4.3.* A $1/8$-inequality $\xi^{8\text{-}(a_m)}$ is valid if there is an integer solution $(a_m)$ satisfying

1. Complementarities: $a_1 + a_8 = n+1$, $a_2 + a_7 = n+1$, $a_3 + a_6 = n+1$, $a_4 + a_5 = n+1$
2. Non-decreasing: $2 \le a_1 \le a_2 \le a_3 \le a_4 \le (n+1)/2$
3. Sub-additivities: $2a_1 \ge a_2$, $a_1 + a_2 \ge a_3$, $a_1 + a_3 \ge a_4$, $2a_2 \ge a_4$; $a_1 + 2a_4 \ge n+1$, $a_2 + a_3 + a_4 \ge n+1$, $3a_3 \ge n+1$,
4. Defining: $x_{i_1-1} + 1 \le a_1 \le x_{i_1}$, $x_{i_2-1} + 1 \le a_2 \le x_{i_2}$, $x_{i_3-1} + 1 \le a_3 \le x_{i_3}$, $x_{i_4-1} + 1 \le a_4 \le x_{i_4}$

where $x_{i_1}, x_{i_2}, x_{i_3}, x_{i_4} \in X$ are the smallest indices $x \in X$ of $\xi_x \ge 1/8, 2/8, 3/8, 4/8$.

# 5 Worst case analysis

In the shooting experiment, we have observed a spike of the size of $1/k$-facets at $k = n$. In this section, we define the strength of a facet and analyze bounds on the strength of the $1/n$-facets introduced in Section 2.2.

## 5.1 *Measures of facet size and strength*

Although the facets of a polyhedron are all necessary elements of its description as a system of inequalities, there are ways in which some facets could be considered more important than others. One measurement for facet importance is the result of a shooting experiment, as described in Section 3. From the perspective of the origin (or an alternative point used as the origin of the shots), the shooting experiment measures the percent of the field of vision occupied by the individual facets; as this can be thought of as a measure of size from a given perspective for the remainder of the paper we will call facets that were frequently hit in the shooting experiment *large* and those that are not frequently hit *small*.

An alternative view of facet importance is known as the worst case analysis which for an individual facet $\xi t \leq 1$ considers the problem of maximizing the objective $\xi t$ over the polyhedron after removal of the facet $\xi t \leq 1$. Since a facet is necessary for the description, the optimal objective value is always larger than one; however, if it is very close to one we may consider such facet to be *weak*, and if this value is larger we consider the facet to be *strong*.

Shim, Cao, and Chopra [18] have computationally demonstrated that these notions of large and small, and strong and weak go together in the sense that these measures are highly correlated with each other, i.e. large facets were consistently found to be strong, and small facets were found to be weak. Since computing the strength or weakness of all individual facets is computationally intensive as $n$ and the number of facets grow, the worst case analysis was only performed for values of $n \leq 26$. Their work is an important confirmation of the validity of the shooting experiment as an accurate measurement of facet importance for the master knapsack polytope; previously some had questioned the validity of shooting experiments because the result depends on the selection of a point from which to fire off the shots (the origin in our case). This confirmation is good news for the computational evaluation of facets because the shooting experiment is much more computationally tractable than the worst case analysis of all individual facets. In the following we define a slightly more general notion of strength and prove that it is equivalent to the notion of strength informally described above and used in [18].

## 5.2 Gradient lemma

Let $z^{IP}(v)$ denote the optimal value of the knapsack problem with objective $v \geq 0$ and let $z^{LP(\xi)}(v)$ denote the optimal value of the LP problem over the system of the knapsack equation, the non-negativity constraints and all knapsack facets except $\xi t \leq 1$.

Consider the primal linear programming problem over the system (20) describing $P(K(n))$ with $\xi^1 t \leq 1$ deleted

$$\begin{aligned} \max\ & vt \\ st\ & lin(n) \cdot t = 1 \\ & \xi^1 t \leq 1 \\ & \vdots \\ & \xi^K t \leq 1 \\ & t \geq 0, \end{aligned} \tag{20}$$

where $lin(n) = (1/n, 2/n, \ldots, n/n)$ is the coefficient vector of the equation replacing the knapsack equation. Then, $z^{LP(\xi^1)}(v)$ is the optimal solution. Note that the dual problem can be written as

$$\min\ x_0 + x_1 + \ldots + x_K, \tag{21}$$

$$st\ \ x_0 \cdot lin(n) + x_1 \xi^1 + \ldots + x_K \xi^K - y = v, \tag{22}$$

$$x_1, \ldots, x_K \text{ and } y \text{ are all nonnegative.} \tag{23}$$

We fix $x_1 = 0$ when $\xi^1 t \leq 1$ is deleted.

The *LP-relaxation gap* of $\xi t \leq 1$ is defined to be

$$\max_{v \geq 0} \frac{z^{LP(\xi)}(v)}{z^{IP}(v)}.$$

Note that its appropriate to use max instead of sup above as $z^{LP(\xi)}(v)$ will always be bounded and have an optimal solution; this follows from the fact that any solution $t$ satisfies $t \geq 0$ and $lin(n) \cdot t = 1$. A knapsack facet $\xi$ is said to be *strong* if the gap is large and *weak* otherwise. In this section, we show that the gap is same as

$$z^{LP(\xi)}(\xi).$$

**Lemma 5.1.** *Let $\xi^k t \leq 1, k = 1, \ldots, K$, be the knapsack facets of $P(K(n))$ with $\xi_1^k = 0$ and $\xi_n^k = 1$. Then, for each $k = 1, \ldots, K$, the LP-relaxation gap of $\xi^k$ is*

$$\max_{v \geq 0} \frac{z^{LP(\xi^k)}(v)}{z^{IP}(v)} = \frac{z^{LP(\xi^k)}(\xi^k)}{z^{IP}(\xi^k)} = z^{LP(\xi^k)}(\xi^k). \tag{24}$$

*Proof.* Since $\xi_1^k = 0$ for all $k = 1, \ldots, K$, the first component of the equation in (22) implies

$$x_0 = (v_1 + y_1)n \geq 0 \tag{25}$$

which can be added to (23) having all dual variables non-negative.

Let $t^{IP}$ be an optimal solution to the primal problem (20) and let $(x^{IP}, y^{IP})$ be an optimal solution to the dual problem described in (21)–(23). Then, we have strong duality

$$vt^{IP} = \sum_{i=0}^{K} x_i^{IP}$$

and we may assume by scaling $v$ that

$$z^{IP}(v) = vt^{IP} = \sum_{i=0}^{K} x_i^{IP} = 1. \tag{26}$$

Since all $x_i^{IP}$ are non-negative in (23) and (25), equality (26) is followed by

$$x_i^{IP} \leq 1 \text{ for } i = 0, \ldots, K. \tag{27}$$

We only need to show the theorem for $k = 1$. If $\xi^1 t^{IP} < 1$, then $x_1^{IP} = 0$ by complementary slackness and eliminating $\xi^1 t \leq 1$ does not change the dual optimal value. We assume that $\xi^1 t^{IP} = 1$. Let $t^{LP}$ be an optimal solution to the primal problem (20) with $\xi^1 t \leq 1$ eliminated. Then, we may assume that

$$\xi^1 t^{LP} \geq 1; \tag{28}$$

otherwise, $t^{LP}$ is feasible for the original (20) and can be switched to $t^{LP} = t^{IP}$.

We complete the proof of the theorem by showing $\xi^1 t^{LP} \geq vt^{LP}$. From (27), (28) and (26), it follows that

$$\begin{aligned} \xi^1 t^{LP} &= (x_1^{IP} + (1 - x_1^{IP}))\xi^1 t^{LP} = x_1^{IP}\xi^1 t^{LP} + (1 - x_1^{IP})\xi^1 t^{LP} \\ &\geq x_1^{IP}\xi^1 t^{LP} + (1 - x_1^{IP})(1) = x_1^{IP}\xi^1 t^{LP} + \left(x_0^{IP} + \sum_{i=2}^{K} x_i^{IP}\right). \end{aligned} \tag{29}$$

Since $lin(n) \cdot t^{LP} = 1$ and $\xi^i t^{LP} \leq 1$ for $i = 2, \ldots, K$, (29) is followed by

$$
\begin{aligned}
\xi^1 t^{LP} &\geq x_1^{IP} \xi^1 t^{LP} + x_0^{IP} + \sum_{i=2}^{K} x_i^{IP} \\
&\geq x_1^{IP} \xi^1 t^{LP} + x_0^{IP} (lin(n) \cdot t^{LP}) + \sum_{i=2}^{K} x_i^{IP} (\xi^i t^{LP}) \\
&= x_0^{IP} (lin(n) \cdot t^{LP}) + x_1^{IP} \xi^1 t^{LP} + \sum_{i=2}^{K} x_i^{IP} (\xi^i t^{LP}) \\
&= x_0^{IP} (lin(n) \cdot t^{LP}) + \sum_{i=1}^{K} x_i^{IP} \xi^i t^{LP} \\
&\geq x_0^{IP} (lin(n) \cdot t^{LP}) + \sum_{i=1}^{K} x_i^{IP} \xi^i t^{LP} - y^{IP} t^{LP} \\
&= \left( x_0^{IP} (lin(n)) + x_1^{IP} \xi^1 + \ldots + x_K^{IP} \xi^K - y^{IP} \right) \cdot t^{LP} = v t^{LP},
\end{aligned}
$$

completing the proof. □

## 5.3 The LP-relaxation gap of a $1/n$-facet

We now present bounds on the LP-relaxation gap of two classes of $1/n$-facets introduced in Section 2.2.

**Theorem 5.2.** *The LP-relaxation gap of $\xi^{n\text{-}(a_m)}$ given by $a_1 = \ldots = a_q = q$ and by $a_i = \lceil i \rceil$ for $q \leq i \leq n-q$ where $1 < q \leq \frac{n}{4}$ satisfies*

$$
z^{LP(\xi^{n\text{-}(a_m)})}(\xi^{n\text{-}(a_m)}) < 1 + \frac{q}{n} \leq 1 + \frac{1}{4}.
$$

*Proof.* Note that

$$
\xi_i^{n\text{-}(a_m)} = \begin{cases} 0 & \text{for } i < q, \\ i/n & \text{for } q \leq i \leq n-q, \text{and} \\ 1 & \text{for } i > n-q. \end{cases}
$$

Let $\hat{\xi}$ be the 1-facet with the shortest half landing. It holds that

$$
\xi^{n\text{-}(a_m)} \leq lin(n) + \frac{q-1}{n} \cdot \hat{\xi}
$$

because for $i \leq n-q$,

$$\xi_i^{n\text{-}(a_m)} - lin(n)_i = \xi_i^{n\text{-}(a_m)} - \frac{i}{n} \leq \frac{i}{n} - \frac{i}{n} = 0 \leq \frac{q-1}{n} \cdot \hat{\xi}_i,$$

and for $i \geq n-q+1$,

$$\xi_i^{n\text{-}(a_m)} - lin(n)_i \leq 1 - \frac{i}{n} \leq 1 - \frac{n-q+1}{n} = \frac{q-1}{n} = \frac{q-1}{n} \cdot \hat{\xi}_i.$$

From $q \leq \frac{n}{4}$, it follows that

$$z^{LP(\xi^{n\text{-}(a_m)})}(\xi^{n\text{-}(a_m)}) \leq 1 + \frac{q-1}{n} < 1 + \frac{q}{n} \leq 1 + \frac{1}{4}.$$

□

**Theorem 5.3.** *The LP-relaxation gap of $\xi^{n\text{-}(a_m)}$ given by $a_i = i$ for $i < q$ even and $a_i = i+1$ for $i < q$ odd and by $a_i = i$ for $q \leq i \leq n-q$ where $q \leq \frac{n}{2}$ is even and $q \leq \frac{n-2}{3}$ if $n$ is odd satisfies*

$$z^{LP(\xi^{n\text{-}(a_m)})}(\xi^{n\text{-}(a_m)}) \leq \frac{n-q+2}{n-q+1} = 1 + \frac{1}{n-q+1} \leq 1 + \frac{2}{n+2}.$$

*Proof.* With $\alpha = \frac{n-q+2}{n-q+1}$, it holds that

$$\alpha \cdot lin(n)_i = \alpha \cdot \frac{i}{n} \geq \frac{i}{n} + \frac{1}{n}$$

specially for $i = n-q+1$. Therefore,

$$\alpha \cdot lin(n) = \frac{n-q+2}{n-q+1} \cdot lin(n) \geq \xi^{n\text{-}(a_m)}.$$

From $q \leq n/2$, it is followed by

$$\frac{n-q+2}{n-q+1} = 1 + \frac{1}{n-q+1} \leq 1 + \frac{2}{n+2}.$$

□

The $1/n$-facets analyzed in Theorem 5.2 appear frequently in our shooting experiment. So, although the upper bound on their strength established in Theorem 5.2 does not guarantee that large or strong facets of this category exist for all $n$, it is consistent with our shooting experiment.

The upper bound of the LP-relaxation gap of $1/n$-facets in Theorem 5.3 proves that this second category of $1/n$-facets are weak asymptotically, because the right-hand side of the bound approaches one as $n$ approaches infinity. This result is consistent with our shooting experiment, where facets in this category were seldom observed and were not identified as large. Together, these results give us a deeper understanding of which $1/n$-facets are important.

## 6 Remarks

We have discussed separation and enumeration of $1/k$-facets for knapsack subproblems. We would hypothesize that these facets will be especially helpful on knapsack subproblems where the included indices $L$ are somewhat evenly distributed among the values $\{1,2,\ldots,n\}$. However, there are other cases in which there is reason to believe that the $1/k$-facets will be unhelpful. For example, if most or all of the indices $L$ take relatively small values, then the $1/k$-facets will have some disadvantages for small values of $k$; this is because the superadditivity constraints imply that $\xi_i = 0$ for all $i \le \frac{n}{k+1}$ if $\xi$ is a $1/k$-facet. Thus, the coefficients of the projected facet might be mostly, or entirely zero. Thankfully, in such extreme cases other techniques are likely to be effective, as we now explain.

**Cyclic group relaxation.** Gilmore and Gomory [7] observed a cyclic group repetition for the knapsack subproblems with larger $n$. In particular, if the right-hand side of the knapsack subproblem is considerably larger than all of the indices in the set $L$, then the cyclic group relaxation leads to especially useful inequalities.

**Super-increasing knapsack.** Super-increasing knapsack problems are a class of knapsack sub-problems where $1/k$-facets with small $k$ are not likely to work well. A super-increasing knapsack problem is a knapsack sub-problem (13) with respect to $L = \{i_1 < \ldots < i_l\}$ satisfying

$$\sum_{u<v} i_u \le i_v.$$

A subproblem (13) with $L = \{2^0, 2^1, \ldots, 2^{l-1}\}$ and $n = 2^l - 1$ is an example of super-increasing knapsack problem. In particular, super-increasing knapsack problems have the property that most of the indices in $L$ are concentrated in the smaller values, likely making the $1/k$-facets ineffective for small values of $k$. It will be most interesting to identify strong facets of the convex hull of the integer solutions to a super-increasing knapsack problem. The recent work of Gupta [13] has recently studied super-increasing knapsack problems.

## 7 Conclusion

Our shooting experiment indicates that there is value to focusing on $1/k$-facets for small values of $k$, such as $k$ dividing 6 or 8 when solving the master knapsack problem (and potentially single row relaxations of many integer program). For both the master problem and knapsack subproblems, $1/k$-facets are easier to enumerate and separate for small values of $k$, and for knapsack subproblems the number of projections of $1/k$-facets depends only on the size of the subproblem and not on the size $n$ of the master problem. We have also studied the strength of two classes of $1/n$-facets, which were introduced by Aráoz et al. [2]; in agreement with our shooting experiment, we found that one class is likely strong and the other is asymptotically weak.

## References

1. Aráoz, J.: Polyhedral neopolarities, Ph.D. Thesis, University of Waterloo, Department of Computer Science (1974)
2. Aráoz, J., Evans, L., Gomory, R., Johnson, E.: Cyclic group and knapsack facets. Math. Program. **96**, 377–408 (2003)
3. Basu, A., Hildebrand, R., Köppe, M., Molinaro, M.: A $(k+1)$-slope theorem for the $k$-dimensional infinite group relaxation. SIAM J. Optim. **23(2)**, 1021–1040 (2011)
4. Cornuéjols, G., Molinaro, M.: A 3 slope theorem for the infinite relaxation in the plane. Math. Program. Ser. A **142**, 83–105 (2013)
5. Dash, S., Günlük, O.: Valid inequalities based on the interpolation procedure. Math. Program. **106**, 111–136 (2006)
6. Evans, L.A.: Cyclic group and knapsack facets with applications to cutting planes. Ph.D. Thesis, Georgia Institute of Technology (2002)
7. Gilmore, P.C., Gomory, R.E.: The theory and computation of knapsack functions. Oper. Res. **14**, 1045–1074 (1966)
8. Gomory, R.E.: Some polyhedra related to combinatorial problems. Linear Algebra Appl. **2**, 451–558 (1969)
9. Gomory, R.E., Johnson, E.L.: Some continuous functions related to corner polyhedra. Math. Program. **3**, 23–85 (1972)
10. Gomory, R.E., Johnson, E.L.: Some continuous functions related to corner polyhedra, II. Math. Program. **3**, 359–389 (1972)
11. Gomory, R.E., Johnson, E.L.: T-space and cutting planes. Math. Program. **96**, 341–375 (2003)
12. Gomory, R.E., Johnson, E.L., Evans, L.: Corner polyhedra and their connection with cutting planes. Math. Program. **96**, 321–339 (2003)
13. Gupte, A.: Convex hulls of superincreasing knapsacks and lexicographic orderings. Manuscript. http://arxiv.org/abs/1503.03742 (2015)
14. Hunsaker, B.: Measuring facets of polyhedra to predict usefulness in branch-and-cut algorithms. Ph.D. Thesis, Georgia Institute of Technology (2003)
15. Kuhn, H. W.: Discussion. In: Proceedings of the IBM Scientific Symposium on Combinatorial Problems, 16–18 March 1964, IBM Data Processing Division, pp. 118–121. White Plains, New York (1966)
16. Shim, S.: Large scale group network optimization. Ph.D. Thesis, Georgia Institute of Technology (2009)
17. Shim, S., Johnson, E.L.: Cyclic group blocking polyhedra. Math. Program. **138**, 273–307 (2013)
18. Shim, S., Chopra, S., Cao, W.: The worst case analysis of strong knapsack facets. Manuscript. www.researchgate.net/publication/263181439 (2014)
19. Shu, Y., Chopra, S., Johnson, E.L., Shim, S.: Binary group facets with complete support and non-binary coefficients. Oper. Res. Lett. **41**, 679–684 (2013)

# On the Performance of SQP Methods for Nonlinear Optimization

**Philip E. Gill, Michael A. Saunders, and Elizabeth Wong**

**Abstract** This paper concerns some practical issues associated with the formulation of sequential quadratic programming (SQP) methods for large-scale nonlinear optimization. SQP methods find approximate solutions of a sequence of quadratic programming (QP) subproblems in which a quadratic model of the Lagrangian is minimized subject to the linearized constraints. Numerical results are given for 1153 problems from the `CUTEst` test collection. The results indicate that SQP methods based on maintaining a quasi-Newton approximation to the Hessian of the Lagrangian function are both reliable and efficient for general large-scale optimization problems. In particular, the results show that in some situations, quasi-Newton SQP methods are more efficient than interior methods that utilize the exact Hessian of the Lagrangian. The paper concludes with discussion of an SQP method that employs both approximate and exact Hessian information. In this approach the quadratic programming subproblem is either the conventional subproblem defined in terms of a positive-definite quasi-Newton approximate Hessian or a convexified subproblem based on the exact Hessian.

**Keywords** Nonlinear programming • Sequential quadratic programming • SQP methods • Interior methods • Convexification • Active-set methods • Quadratic programming

**Mathematics subject classification (2010):** 49J20, 49J15, 49M37, 49D37, 65F05, 65K05, 90C26, 90C30, 90C51, 90C55

---

P.E. Gill (✉) • E. Wong
Department of Mathematics, UC San Diego, La Jolla, CA 92093-0112, USA
e-mail: pgill@ucsd.edu; elwong@ucsd.edu

M.A. Saunders
Systems Optimization Laboratory, Department of Management Science and Engineering, Stanford University, Stanford, CA 94305-4121, USA
e-mail: saunders@stanford.edu



B. Defourny, T. Terlaky (eds.), *Modeling and Optimization: Theory and Applications*, Springer Proceedings in Mathematics & Statistics 147,
DOI 10.1007/978-3-319-23699-5_5

## 1 Introduction

This paper concerns the formulation of a sequential quadratic programming (SQP) method for the solution of the nonlinear optimization problem

$$\begin{array}{ll} \underset{x\in\mathbb{R}^n}{\text{minimize}} & f(x) \\ \text{subject to} & \ell \le \begin{pmatrix} x \\ Ax \\ c(x) \end{pmatrix} \le u, \end{array} \tag{1}$$

where $f(x)$ is a linear or nonlinear objective function, $c(x)$ is a vector of $m$ nonlinear constraint functions $c_i(x)$, $A$ is a matrix, and $\ell$ and $u$ are vectors of lower and upper bounds. For simplicity, in our discussion of the theoretical aspects of SQP methods we assume that the problem has the form

$$\text{(NP)} \qquad \underset{x\in\mathbb{R}^n}{\text{minimize}}\; f(x) \quad \text{subject to} \quad c(x) \ge 0,$$

where $f$ and the $m$ components of the constraint vector $c$ are assumed to be twice continuously differentiable for all $x \in \mathbb{R}^n$. Any linear constraints and simple bound constraints are included in the definition of $c$. However, we emphasize that the exploitation of the properties of linear constraints is an important issue in the solution of large-scale problems.

No assumptions are made about $f$ and $c$, other than twice differentiability; for example, the problem need not be convex. The vector $g(x)$ denotes the gradient of $f$ evaluated at $x$, and $J(x)$ denotes the $m \times n$ constraint Jacobian, which has $i$th row $\nabla c_i(x)^T$, the gradient of the $i$th constraint function $c_i$ evaluated at $x$. The Lagrangian associated with (NP) is $L(x, y) = f(x) - c(x)^T y$, where $y$ is the $m$-vector of dual variables associated with the inequality constraints $c(x) \ge 0$. The Hessian of the Lagrangian with respect to $x$ is denoted by $H(x, y) = \nabla^2 f(x) - \sum_{i=1}^m y_i \nabla^2 c_i(x)$.

Sequential quadratic programming methods find approximate solutions of a sequence of quadratic programming (QP) subproblems in which a quadratic model of the Lagrangian function is minimized subject to the linearized constraints. In a merit-function based SQP method, the QP solution provides a direction of improvement for a function that represents a compromise between the (often conflicting) aims of minimizing the objective function and reducing the constraint violations. Many SQP methods use an active-set quadratic programming method to solve the QP subproblem. In this situation, the SQP method has a major/minor iteration structure in which each minor iteration is an iteration of the active-set QP solver. The work for a minor iteration is dominated by the cost of solving a system of symmetric indefinite linear equations defined in terms of a subset of the variables and constraints.

Interior-point (IP) methods use a completely different approach to handle the inequality constraints of problem (NP). Interior methods follow a continuous path that terminates at a solution of (NP). In the simplest case, the path is parameterized by a positive scalar parameter $\mu$ that may be interpreted as a perturbation for the first-order optimality conditions for the problem (NP). If $x(\mu)$ denotes a point on the path associated with the parameter value $\mu$, then $x(0) = x^*$, where $x^*$ is a solution of (NP). Each point on the path may be found by applying Newton's method to a system of nonlinear equations that represents perturbed optimality conditions for the original problem (NP). Each iteration of Newton's method involves a system of linear equations defined in terms of the derivatives of $f$ and $c$. The Newton equations may be written in symmetric form, in which case each iteration requires the solution of a single symmetric indefinite system of equations involving the derivatives of every constraint in the problem.

The conventional wisdom is that when solving a general nonlinear problem "from scratch" (i.e., with no prior knowledge of the properties of a solution), software based on an IP method is generally faster and more reliable than software based on an SQP method. However, as SQP methods have the potential to capitalize on a good initial starting point, they are considered to be more effective for solving a sequence of similar problems, such as a sequence of discretized continuous problems for which some underlying discretization is being refined. This claim is difficult to verify, however, as most test collections include unrelated problems of varying sizes and difficulty, or groups of problems with similar characteristics but slightly different formulations. Providing a fair comparison of SQP and IP methods is also complicated by the fact that very few SQP software packages are able to exploit the second derivatives of a problem. (This issue is considered further in Section 4.) Moreover, IP methods are more straightforward to implement with second derivatives, and most software test environments for optimization provide test problems for which second derivatives are available automatically. Unfortunately, there are many practical problems for which even *first* derivatives are difficult or expensive to compute. Test results from second-derivative methods are unlikely to be representative in this case.

The purpose of this paper is twofold. First, we provide a comparison of two widely used software packages for general nonlinear optimization, one an IP method (`IPOPT` [45, 47, 48]) and one an SQP method (`SNOPT7` [23]). These packages are applied to almost all the problems from the `CUTEst` testing environment [30]. The tests are formulated so that the same derivative information is provided to both packages. In this environment it is shown that conclusions concerning the relative performance of first-derivative IP and SQP methods are more nuanced than the conventional wisdom. In particular, it is shown that active-set methods can be efficient for the solution of "one-off" problems, as is the case with active-set methods for linear programming. In other words, SQP methods may be best suited for some problems, and IP methods may be best for others.

If software is intended to be used in an environment in which second derivatives are available, then it is clear that the method that can best exploit these derivatives should be used. The second purpose of this paper is to extend conventional SQP methods so that second derivatives can be exploited reliably and efficiently when they are available. These extensions are motivated by some comparisons of first-derivative SQP methods with second-derivative IP methods.

## 2 Background on SQP methods

The two principal ingredients of a merit-function based SQP method are: (i) a scalar-valued merit function $\mathscr{M}$ that provides a measure of the quality of a given point as an estimate of a solution of the constrained problem; and (ii) a direction of improvement for $\mathscr{M}$ defined as the solution of a quadratic programming subproblem. As in the unconstrained case, the merit function is used in conjunction with a line-search model to define a sufficient decrease in $\mathscr{M}$ at each major iteration. Here we focus on the formulation and solution of the QP subproblem. For more background on the properties of SQP methods, see, e.g., Gill and Wong [17].

### 2.1 *Properties of the QP subproblem*

Given the $k$th estimate $(x_k, y_k)$ of the primal and dual solution of (NP), a conventional line-search SQP method defines a direction $p_k = \hat{x}_k - x_k$, where $\hat{x}_k$ is a solution of the QP subproblem

$$\begin{aligned} &\underset{x\in\mathbb{R}^n}{\text{minimize}} && g(x_k)^T(x-x_k) + \tfrac{1}{2}(x-x_k)^T\hat{H}_k(x-x_k) \\ &\text{subject to} && J(x_k)(x-x_k) \geq -c(x_k), \end{aligned} \tag{2}$$

with $\hat{H}_k$ an exact or approximate Hessian of the Lagrangian. If the QP subproblem (2) has a solution, then the QP first-order optimality conditions imply the existence of a primal-dual pair $(\hat{x}_k, \hat{y}_k)$ such that

$$g(x_k) + \hat{H}_k(\hat{x}_k - x_k) = J(x_k)^T\hat{y}_k, \qquad \hat{y}_k \geq 0, \tag{3}$$

$$r(\hat{x}_k)\cdot\hat{y}_k = 0, \qquad r(\hat{x}_k) \geq 0, \tag{4}$$

where $r(x)$ is the vector of constraint residuals $r(x) = c(x_k) + J(x_k)(x - x_k)$, and $a\cdot b$ denotes the vector with $i$th component $a_i b_i$. At any feasible point $x$ for (2), the active set associated with the QP subproblem is given by

$$\mathscr{A}(x) = \{\, i : r_i(x) = [c(x_k) + J(x_k)(x-x_k)]_i = 0 \,\}.$$

The optimality conditions for the QP subproblem (2) may be characterized in terms of an index set $\mathscr{W}_k \subseteq \mathscr{A}(\hat{x}_k)$ such that the rows of $J(x_k)$ with indices in $\mathscr{W}_k$ are linearly independent. If the conditions (3)–(4) hold for at least one primal-dual pair, then there must exist a nonnegative $\hat{y}_k$ and index set $\mathscr{W}_k$ such that $[\hat{y}_k]_i = 0$ for $i \notin \mathscr{W}_k$, and

$$g(x_k) + \hat{H}_k(\hat{x}_k - x_k) = J_w(x_k)^T\hat{y}_w, \qquad \hat{y}_w \geq 0, \tag{5}$$

$$c_w(x_k) + J_w(x_k)(\hat{x}_k - x_k) = 0, \qquad r(\hat{x}_k) \geq 0, \tag{6}$$

where $c_w(x_k)$ and $J_w(x_k)$ denote the rows of $c(x_k)$ and $J(x_k)$ associated with indices in $\mathcal{W}_k$, and $\hat{y}_w$ is the subvector of $\hat{y}_k$ associated with the indices in $\mathcal{W}_k$. The linear equalities associated with the conditions (5)–(6) may be written in matrix form

$$\begin{pmatrix} \hat{H}_k & J_w(x_k)^T \\ J_w(x_k) & 0 \end{pmatrix} \begin{pmatrix} p_k \\ -\hat{y}_w \end{pmatrix} = - \begin{pmatrix} g(x_k) \\ c_w(x_k) \end{pmatrix}, \tag{7}$$

where $p_k = \hat{x}_k - x_k$. The index set $\mathcal{W}_k$ is said to be *second-order consistent* with respect to $\hat{H}_k$ if the reduced Hessian $Z_w^T \hat{H}_k Z_w$ is positive definite, where the columns of $Z_w$ form a basis for the null-space of $J_w(x_k)$. If $\mathcal{W}_k$ is second-order consistent with respect to $\hat{H}_k$, then the system of equations (7) is nonsingular and defines unique vectors $p_k$ and $\hat{y}_w$ satisfying

$$p_k^T \hat{H}_k p_k = -\big(g(x_k) - J_w(x_k)^T \hat{y}_w\big)^T p_k = -g_L(x_k, \hat{y}_k)^T p_k, \tag{8}$$

where $g_L(x, y)$ denotes the gradient of the Lagrangian function with respect to $x$, i.e., $g_L(x, y) = g(x) - J(x)^T y$. This identity implies that if $\hat{H}_k$ is positive definite, then $p_k$ is a descent direction for the Lagrangian function defined with the QP multipliers.

The dimension of the reduced Hessian $Z_w^T \hat{H}_k Z_w$ can have a crucial influence on the efficiency of an SQP method. This quantity is an estimate of the number of degrees of freedom in problem (NP), i.e., the dimension of the underlying unconstrained problem defined by restricting $f$ to the surface of the active constraints at a solution of the nonlinear problem.

## 2.2 *Active-set methods for the QP subproblem*

The SQP methods discussed in this paper exploit certain benefits derived from using a primal-feasible active-set method to solve the QP subproblem. Primal active-set QP methods have two phases: in phase 1, a feasible point is found by minimizing the sum of infeasibilities; in phase 2, the quadratic objective function is minimized while maintaining feasibility. At each QP iterate, a *working set* of QP constraints is known for which the constraint gradients are linearly independent. At a solution of the QP, this working set is the index set $\mathcal{W}_k$ associated with the QP optimality conditions (5)–(6). At each QP iterate, the working set defines the constraints of an equality constrained subproblem (EQP) whose solution satisfies a system of equations of the form (7). (The precise definition of the working set varies with the method. Some methods restrict the working set to be a subset of the active constraints, while others allow some constraints in the working-set to be strictly satisfied.) If the final QP working set is used to define the initial working set for the next QP subproblem, it is typical for the later QP subproblems to reach optimality in a single iteration because the QP optimality conditions (5)–(6) are satisfied by the solution of the first EQP subproblem, i.e., the EQP solution satisfies the system (7).

## 3 Numerical results

Before addressing the use of second derivatives in SQP methods, we present numerical results that have motivated our work. This section includes a summary of the results obtained by running the SQP package SNOPT7 [23] and the IP package IPOPT [45, 47, 48] on problems in the CUTEst test collection [30]. IPOPT is arguably the most successful and widely used package for nonlinearly constrained optimization. The version of IPOPT used in the runs was version 3.11.8, compiled with the linear solver MA57.

SNOPT7 version 7.4 is a Fortran implementation of the general sequential quadratic programming method discussed in Section 2. SNOPT7 is designed to solve large-scale problems of the form (1). Internally, SNOPT7 transforms this problem into standard form by introducing a vector of slack variables $s$. The equivalent problem is

$$\underset{x,s}{\text{minimize}}\ f(x) \quad \text{subject to} \quad \begin{pmatrix} Ax \\ c(x) \end{pmatrix} - s = 0, \quad l \le \begin{pmatrix} x \\ s \end{pmatrix} \le u. \tag{9}$$

All runs were made on a MacPro configured with a 2.7GHz 12-core Intel Xeon E5 processor and 64GB of RAM. Both IPOPT and SNOPT7 were compiled using gfortran 4.6 with full code optimization and the optimized BLAS library in the Accelerate framework from Apple. The floating-point precision was $2.22 \times 10^{-16}$.

### 3.1 *The active-set method of* SNOPT7

To solve the QP subproblems, SNOPT7 employs the convex QP solver SQOPT [24], which is an implementation of a reduced-Hessian, reduced-gradient active-set method. For a QP subproblem associated with a problem expressed in the form (9), all the inequality constraints are simple bounds. Let $\hat{H}$, $\hat{g}$, and $\hat{A}$ denote the Hessian, gradient, and general constraint matrix associated with the $k$th QP subproblem. In a reduced-gradient method, the general QP constraints $\hat{A}x - s = 0$ are partitioned into the form $Bx_B + Sx_S + Nx_N = 0$, where the matrix $B$ is square and nonsingular, and the matrices $S$, $N$ are the remaining columns of $\begin{pmatrix} \hat{A} & -I \end{pmatrix}$. The vectors $x_B$, $x_S$, $x_N$ are the associated basic, superbasic, and nonbasic components of $(x, s)$ (see Gill, Murray and Saunders [23]). The reduced Hessian $Z^T\hat{H}Z$ is defined in terms of the matrix $Z$ such that

$$Z = P\begin{pmatrix} -B^{-1}S \\ I \\ 0 \end{pmatrix}, \tag{10}$$

where $P$ permutes the columns of $\begin{pmatrix} \hat{A} & -I \end{pmatrix}$ into the order $\begin{pmatrix} B & S & N \end{pmatrix}$. The matrix $Z$ is used only as an operator, i.e., it is not stored explicitly. Products of the form $Zv$ or $Z^T\hat{g}$ are obtained by solving with $B$ or $B^T$. The package LUSOL [21] is used to maintain sparse LU factors of $B$ as the $BSN$ partition changes.

At each QP-iteration, the reduced Hessian is positive semidefinite with at most one zero eigenvalue. If the reduced Hessian is nonsingular, a direction in the superbasic variables is computed from the system

$$Z^T\hat{H}Zp_S = -Z^T\hat{g} \tag{11}$$

using a dense Cholesky factor of $Z^T\hat{H}Z$. If the reduced Hessian is singular, the Cholesky factor is used to define $p_S$ such that $Z^T\hat{H}Zp_S = 0$ and $p_S^T Z^T\hat{g} < 0$. In this implementation, the number of degrees of freedom associated with the QP subproblem is the number of superbasic variables. If this number is large, then solving the reduced Hessian equations (11) dominates the cost of a QP iteration.

## 3.2 *The* CUTEst *test collection*

The CUTEst distribution of January 14, 2015 (Subversion revision 245) contains 1156 problems in standard interface format (SIF). A list of CUTEst problem types and their frequency is given in Table 1. Although many problems allow for the number of variables and constraints to be adjusted in the SIF file, all the tests are based on the default problem dimensions set in the CUTEst distribution. The three problems *recipe*, *s365*, and *s365mod* are omitted because of the potential for a floating-point exception when the problem functions are evaluated at feasible points. The remaining problems form a grand total of 1153 problems ranging in size from *hs1* (two variables and no constraints) to *bdry2* (251001 variables and 250498 constraints). Of these 1153 problems attempted, 137 have more than 3000 degrees of freedom, with the largest nonlinearly constrained problem (*jannson3*) having almost 20000 degrees of freedom at the solution.

**Table 1** The 1156 CUTEst problems listed by frequency and type

| Type | Frequency | Characteristics |
|---|---|---|
| LP | 26 | Linear objective, linear constraints |
| QP | 238 | Quadratic objective, linear constraints |
| UC | 173 | Nonlinear objective, no constraints |
| BC | 141 | Nonlinear objective, bound constraints |
| LC | 70 | Nonlinear objective, linear constraints |
| NC | 408 | Nonlinear objective, nonlinear constraints |
| FP | 100 | Constant objective function |
| NS | 19 | Non-smooth |

The 19 problems *bigbank, bridgend, britgas, concon, core1, core2, gridgena, hs67, hs85, hs87, mconcon, net1, net2, net3, net4, stancmin, twiribg1, twirimd1,* and *twirism1* are non-smooth, but are included in the test-set nevertheless. The 11 problems *fletcbv3, fletchbv, gridgena, indef, lukvle2, mesh, ncvxbqp1, ncvxbqp2, qrtquad, static3,* and *lukvli4* are known to have an objective function that is unbounded below in the feasible region.

Many of the problems are either infeasible or have no known feasible point. The 15 problems *a2nndnil, a5nndnil, arglale, arglble, arglcle, flosp2hh, flosp2hl, flosp2hm, ktmodel, lincont, model, nash, synpop24, toysarah,* and *woodsne* have infeasible linear constraints. For nonlinear constraints, no local optimization method is guaranteed to find a feasible point unless certain restrictions are imposed on the class of constraint functions. In the nonlinear case, the failure of an algorithm to find a feasible point does not imply that the problem is infeasible. In the CUTEst test set, the nonlinear problem *burkehan* is known to be infeasible. Another 14 problems have no known feasible point: *argauss, arwhdne, cont6-qq, drcavty3, eigenb, growth, himmelbd, junkturn, lewispol, lubrif, lubrifc, nuffield, nystrom5, tro41x9.* We conjecture that these problems are infeasible. (Problems *junkturn* and *nystrom5* are feasible if the constraints are perturbed by $10^{-3}$.)

## 3.3 User-specified options

Both `SNOPT7` and `IPOPT` allow the user to replace the values of certain default run-time options. With the exception of an 1800-second time-limit, all `IPOPT` runs were made using the default options. Figure 1 lists the `SNOPT7` and `IPOPT` options that differ from their default values. (For a complete list of these options, see [48] and [23].) `Reduced Hessian dimension` specifies the maximum size of the dense reduced Hessian available for `SQOPT`. If the number of degrees of freedom exceeds this value during the QP solution, `SQOPT` solves a perturbed version of (11) using the conjugate-gradient solver `SYMMLQ` [38]. The default `Major optimality` tolerance for `SNOPT7` is $2 \times 10^{-6}$ (see Section 2.11 of Gill, Murray and Saunders [23]) The larger value of $1.22 \times 10^{-4}$ was used to match the default optimality tolerance of `IPOPT`.

## 3.4 Results obtained using first derivatives only

Table 2 summarizes the results of running `SNOPT7` and `IPOPT` with the `L-BFGS` option on the 1153 test problems. The constraint format (9) allows `SNOPT7` to find a feasible point for the linear constraints before evaluating the objective function and nonlinear constraints. If the linear constraints are infeasible, `SNOPT7` terminates immediately without computing the nonlinear functions. Otherwise, all subsequent major iterates satisfy the linear constraints. (Sometimes this feature helps ensure

**Fig. 1** The SNOPT7 and IPOPT run-time option files

```
BEGIN SNOPT Problem
   Superbasics limit                150000
   Reduced Hessian dimension          4000
   Major iterations                1000000
   Iteration limit                 1000000
   Major optimality tolerance      1.22e-4
   Time limit                         1800
END SNOPT Problem

#BEGIN IPOPT Problem
   max_iter                        1000000
   # next option for L-BFGS only
   hessian_approximation limited-memory
   linear_solver                      ma57
   time_limit                         1800
#END IPOPT Problem
```

**Table 2** SNOPT7 and first-derivative IPOPT on 1153 CUTEst problems

| SNOPT7 | |
|---|---|
| Optimal | 1006 |
| Optimal, but low accuracy | 8 |
| Unbounded | 11 |
| Infeasible constraints | 16 |
| Locally infeasible constraints | 16 |
| Total successes | 1057 |
| False infeasibility | 25 |
| Iteration limit | 4 |
| Time limit | 45 |
| Numerical problems | 13 |
| Final point cannot be improved | 9 |
| Total failures | 96 |

| IPOPT (first derivatives) | |
|---|---|
| Optimal | 772 |
| Optimal, but low accuracy | 161 |
| Unbounded | 3 |
| Infeasible constraints | 10 |
| Locally infeasible constraints | 3 |
| Total successes | 949 |
| False infeasibility | 9 |
| Iteration limit | 37 |
| Time limit | 40 |
| Diverges | 8 |
| Restoration failed | 70 |
| Too few degrees of freedom | 33 |
| Regularization too large | 1 |
| Invalid derivative entry | 5 |
| Segmentation fault | 1 |
| Total failures | 204 |

that the functions and gradients are only computed at points where they are well defined.) For brevity, Table 2 lists the number of infeasible problems found by SNOPT7 without distinguishing between linear and nonlinear constraints.

Of the 19 nonsmooth problems, SNOPT7 solved all but *hs87* and *net4*. IPOPT solved all but *bigbank*, *gridgena*, *hs87*, and *net4*. All 11 unbounded problems were identified correctly by SNOPT7. IPOPT terminated 11 cases with an "unbounded or diverging" diagnostic message, with *indef*, *mesh*, and *static3* being correctly identified as unbounded. Problem *gausselm* terminated with a segmentation fault after IPOPT failed to allocate sufficient memory for MA57.

If any QP subproblem is infeasible, or the Lagrange multipliers of the subproblem become large, then SNOPT7 switches to "elastic mode." In this mode, the nonlinear constraint functions are allowed to violate their bounds by an amount that is multiplied by a positive weight and included in the objective function (see, e.g., Gill, Murray and Saunders [23]). This feature allows SNOPT7 to find a local minimizer of the sum of infeasibilities if the nonlinear constraints appear to be infeasible. As mentioned above, the calculation of such a point does not necessarily imply the problem is infeasible. A run was considered to have "failed" if a final point of local infeasibility was declared for a problem that is known to be feasible. SNOPT7 terminated at a "false" infeasible point for 25 problems: *a4x12*, *broydnbd*, *discs*, *drugdis*, *eigmaxc*, *eigminc*, *flosp2th flt*, *hadamard*, *hatfldf*, *hs61*, *lukvle11*, *lukvle16*, *lukvle17*, *lukvle18*, *mss1*, *mss2*, *mss3*, *optcdeg3*, *powellsq*, *s316-322*, *tro21x5*, *vanderm1*, *vanderm2*, and *vanderm3*. This large number of false infeasibilities provides a somewhat misleading picture of the effectiveness of SNOPT7 for finding a feasible point. In particular, a total of 11 of the "infeasible" problems: *broydnbd*, *discs*, *drugdis*, *eigmaxc*, *eigminc*, *flt*, *hadamard*, *hs61*, *mss2*, *mss3*, and *tro21x5*, solve to optimality with the default optimality tolerance $10^{-6}$. Similarly, the final sum of infeasibilities for the 7 problems *a4x12*, *flosp2th hatfldf*, *lukvle17*, *lukvle18*, *vanderm1*, and *vanderm2*, was of the order of $10^{-5}$. Problems *fletcher* and *lootsma* have feasible solutions, but their initial points are infeasible and stationary for the sum of infeasibilities. In this situation, the initial point satisfies the first-order conditions for a minimizer of the merit function and SNOPT7 terminates immediately. As this study does not recognize a qualitative distinction between a local and global solution, the outcomes for *fletcher* and *lootsma* are listed as successful.

IPOPT with L-BFGS determined that 22 problems are infeasible. Of these, 9 problems: *artif*, *cresc100*, *lippert2*, *lukvle16*, *lukvli17*, *pfit2*, *pfit4*, *powellsq*, and *wachbieg*, are listed as failures because of the existence of known feasible points.

The results are summarized using performance profiles proposed by Dolan and Moré [6]. A performance profile provides an "at-a-glance" comparison of the performance of a set $\mathscr{S}$ of $n_s$ solvers applied to a test set $\mathscr{P}$ of $n_p$ problems. For each solver $s \in \mathscr{S}$ and problem $p \in \mathscr{P}$ in a profile, the number $t_{ps}$ is the performance measure (i.e., the solve-time or number of function evaluations) for solver $s$ on problem $p$. To compare the performance of a problem $p$ over the different solvers, the *performance ratio* for each successfully solved problem and solver is defined as

$$r_{ps} = \frac{t_{ps}}{\min\{t_{ps} : s \in \mathscr{S}\}}.$$

If $r_{ms}$ denotes the maximum time (or function evaluations) needed over all problems that were solved successfully, then the performance ratio for problems that failed is defined as some value greater than $r_{ms}$. Given the set of performance ratios, a function $P_s(\sigma)$ is defined for each solver such that

$$P_s(\sigma) = \frac{1}{n_p}|\{p \in \mathscr{P} : r_{ps} \leq \sigma\}|,$$

where $\sigma \in [1, r_{ms}]$. The value $P_s(\sigma)$ is the fraction of problems for solver $s$ that were solved within $\sigma$ of the best time. $P_s(1)$ is the fraction of problems for which $s$ was the fastest solver. The value $P_s(r_{ms})$ gives the fraction of problems solved successfully by solver $s$. The presented performance profiles are log-scaled, with $\tau = \log_2(\sigma)$ on the $x$-axis and the function

$$P_s(\tau) = \frac{1}{n_p}|\{p \in \mathscr{P} : \log_2(r_{ps}) \leq \tau\}|,$$

on the $y$-axis for each solver. The $y$-axis can be interpreted as the fraction of problems that were solved within $2^\tau$ of the best time. Because the $y$-axis is the fraction of problems solved, and the $x$-axis is the factor of time needed to solve a problem, the "best" solver should have a function $P_s(\tau)$ that lies towards the upper-left of the graph. Recorded solve times of less than 0.001 seconds are replaced by 0.001 to prevent division by zero in the calculation of the performance ratios.

Figure 2 gives the performance profiles for the solve times (in seconds) and total number function evaluations required by SNOPT7 and IPOPT on all 1153 problems. The left figure profiles the solve times, the right figure profiles function evaluations. In these profiles, an algorithm is considered to have solved a problem successfully if one of the first five outcomes listed in Table 2 occurs.

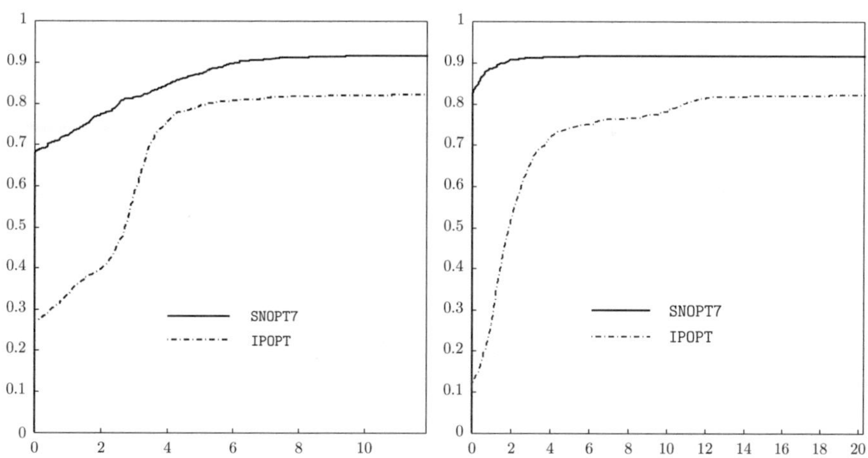


**Fig. 2** Performance profiles for SNOPT7 and IPOPT on 1153 CUTEst test problems using first derivatives. The IPOPT results were obtained using an L-BFGS approximate Hessian. The left figure profiles solve times (in seconds); the right figure profiles function evaluations

**Table 3** SNOPT7 and second-derivative IPOPT on 1153 CUTEst problems

| SNOPT7 | |
|---|---|
| Optimal | 1006 |
| Optimal, but low accuracy | 8 |
| Unbounded | 11 |
| Infeasible constraints | 16 |
| Locally infeasible constraints | 16 |
| Total successes | 1057 |
| False infeasibility | 25 |
| Iteration limit | 4 |
| Time limit | 45 |
| Numerical problems | 13 |
| Final point cannot be improved | 9 |
| Total failures | 96 |

| IPOPT (second derivatives) | |
|---|---|
| Optimal | 1013 |
| Optimal, but low accuracy | 18 |
| Unbounded | 2 |
| Infeasible constraints | 9 |
| Locally infeasible constraints | 6 |
| Total successes | 1048 |
| False infeasibility | 11 |
| Iteration limit | 3 |
| Time limit | 28 |
| Too few degrees of freedom | 33 |
| Diverges | 5 |
| Restoration failed | 19 |
| Regularization too large | 2 |
| Invalid derivative entry | 2 |
| Search direction too small | 2 |
| Total failures | 105 |

### *3.5 Comparisons with* IPOPT *using second-derivatives*

In this section we consider results obtained by running SNOPT7 with first derivatives (the only option) and IPOPT with the second derivatives (the default option). The results are summarized in Table 3.

The second-derivative version of IPOPT solved all of the 19 nonsmooth problems except *britgas*, *net2*, and *net4*. IPOPT identified 26 problems as being infeasible. Of these, the 11 problems *artif*, *brainpc2*, *cresc100*, *cresc132*, *cresc50*, *lukvle16*, *net4*, *pfit1*, *pfit2*, *powellsq*, and *wachbieg*, have known feasible points and are included in the list of failures. Of the 7 problems that IPOPT identified as being unbounded, only *mesh* and *static3* were feasible with an unbounded objective. The other 5 problems are listed as "diverging" in Table 3.

Figure 3 gives the performance profiles for the solve times and function evaluations required by SNOPT7 (first derivatives only) and second-derivative IPOPT on all 1153 problems. As above, an algorithm is considered to have solved a problem successfully if one of the first five outcomes listed in Table 3 occurs.

Two conclusions may be drawn from these results. First, IPOPT with the second-derivative option is significantly more robust than IPOPT with first derivatives only, which is why the IPOPT documentation strongly recommends the second-derivative option. Second, software based on a first-derivative SQP method can be competitive with software based on an IP method using second derivatives. The remaining discussion focuses upon the source of this competitiveness and how it may be exploited for the development of second-derivative SQP methods.

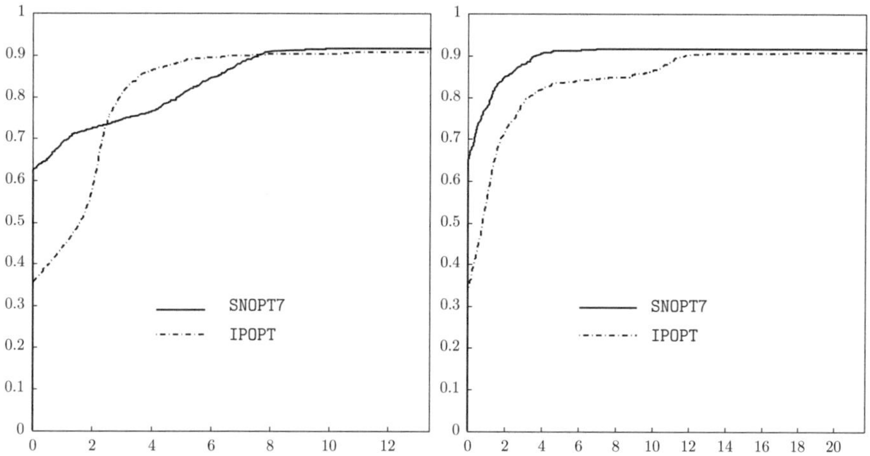


**Fig. 3** Performance profiles for SNOPT7 (first derivatives only) and second-derivative IPOPT on 1153 CUTEst test problems. The left figure profiles the solve times (in seconds), the right figure profiles function evaluations

An important quantity that influences the efficiency of an optimization method is the number of degrees of freedom $n_{df}$ at a solution. The 1153 problems in the CUTEst test collection include 1016 problems with $n_{df} \leq 3000$ and 1095 problems with $n_{df} \leq 4000$[1]. Generally speaking, methods that maintain an explicit reduced Hessian when solving the QP subproblem (such as the QP solver SQOPT used in SNOPT7) become less efficient as the number of degrees of freedom increases. Figure 4 illustrates how the efficiency of SNOPT7 is influenced by the number of degrees of freedom. Figure 4 profiles the solve times for SNOPT7 and IPOPT on the 1016 CUTEst test problems with $n_{df} \leq 3000$. A comparison of the solution-time profiles of Figures 3 and 4 indicates that overall, SNOPT7 is more competitive on problems with few degrees of freedom at the solution. The IPOPT package uses a direct factorization of the KKT matrix, which implies that the efficiency is relatively unrelated to the number of degrees of freedom. It follows that as the number of degrees of freedom increases, the number of problems for which IPOPT has a faster solution time increases. For example, on the 68 problems with $n_{df} > 4000$ only 24 problems are solved faster with SNOPT7. These 68 problems provide a substantial test of the conjugate-gradient linear system solver in SQOPT.

This inefficiency may be removed by using a QP solver that maintains an explicit reduced Hessian when the number of degrees of freedom is small, and uses direct factorization when the number of degrees of freedom is large. The QP-package SQIC [18] implements a method based on this strategy.

[1] Here, the value of $n_{df}$ is taken as the number of degrees of freedom at a solution found by SNOPT7.

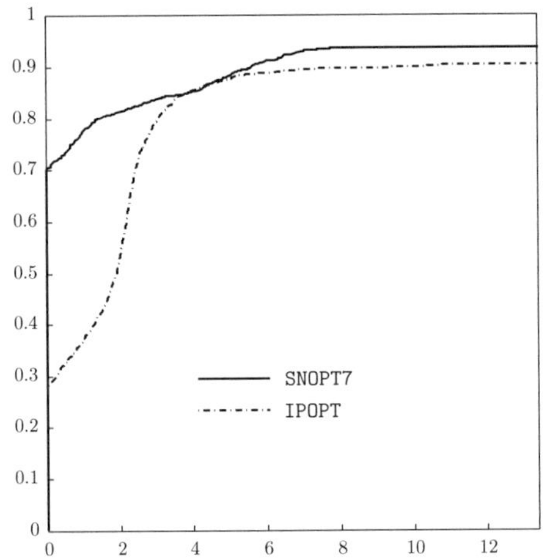


**Fig. 4** Performance profiles of solve times for SNOPT7 (first derivatives only) and second-derivative IPOPT on 1016 CUTEst test problems with no greater that 3000 degrees of freedom

Given an efficient QP solver, once the QP working set settles down, the efficiency of SNOPT7 depends largely on whether or not the limited-memory method is able to adequately represent the Lagrangian Hessian. For example, many of the 68 large problems have no constraints or only simple bounds, and in these cases, the large number of major iterations is consistent with results obtained by other limited-memory quasi-Newton methods (see, e.g., [1, 14]). This is one situation in which the use of second derivatives, when available, can have a significant impact on the rate of convergence of the SQP method. Figure 5, which profiles the solve times for SNOPT7 and IPOPT on all 296 CUTEst test problems with either no constraints or only simple bounds, indicates that, overall, second derivatives provide a significant edge on this class of problem.

## 4 Second-derivative SQP methods

The numerical results of the previous section confirm the widely held view that quasi-Newton SQP methods are effective at providing a good estimate of the indices of the active constraints at a solution of (NP). Once the QP working set settles down, the efficiency of a quasi-Newton SQP method then depends largely on whether or not the limited-memory method is able to adequately represent the Lagrangian Hessian. It is in this situation that the use of second derivatives, when available, can have a significant impact on the rate of convergence of the SQP method.

In this section we outline an SQP method that incorporates both approximate and exact Hessian information. First, the method uses the solution of a convex QP based on a quasi-Newton Hessian to identify an estimate of the working set at a

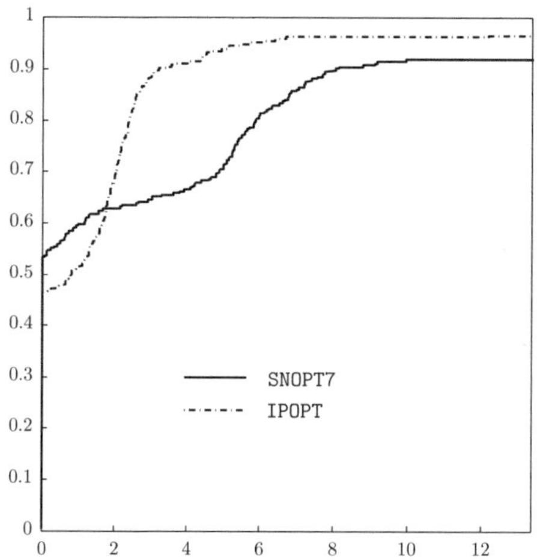


**Fig. 5** Performance profiles of solve times for SNOPT7 (first derivatives only) and second-derivative IPOPT on all 296 CUTEst test problems with either no constraints or only simple bounds

solution. This estimate is used to initiate a sequence of QP subproblems defined by the Hessian of the Lagrangian. The approach is to proceed to solve the nonconvex QP subproblem as in a conventional SQP method. However, as the QP iterations proceed, the QP Hessian is modified implicitly in such a way that the sequence of QP iterates is equivalent to that associated with a related *convex* QP subproblem. This "convexification" process is related to some well-known methods for unconstrained optimization that modify a subproblem "on-the-fly."

Other methods that identify the active set using a convex QP based on a BFGS approximation of the Hessian have been proposed by Byrd et al. [4] and Gould and Robinson [27–29].

## 4.1 Difficulties associated with using second derivatives in SQP

If the Hessian of the Lagrangian $H(x_k, y_k)$ at $(x_k, y_k)$ is positive definite and the QP subproblem Hessian is $\widehat{H}_k = H(x_k, y_k)$, then the SQP search direction $p_k$ satisfies the inequality $g_L(x_k, \hat{y}_k)^T p_k < 0$, and $p_k$ is a descent direction for the Lagrangian defined with multipliers $y = \hat{y}_k$ (see (8)). The curvature condition $p_k^T H(x_k, y_k) p_k > 0$ is sufficient for the existence of a step length that provides a sufficient decrease for several merit functions that have been proposed in the literature; e.g., the $\ell_1$ penalty function (Han [35] and Powell [39]) and various forms of the augmented Lagrangian merit function (Han [35], Schittkowski [40], and Gill, Murray, Saunders and Wright [22]).

If problem (NP) is not convex, the Hessian of the Lagrangian may be indefinite, even in the neighborhood of a solution. This situation creates a number of difficulties in the formulation and analysis of a conventional SQP method.

(i) The QP subproblem (2) may be nonconvex, which implies that the objective of (2) may be unbounded below in the feasible region, and that there may be many local solutions. In addition, nonconvex QP is NP-hard—even for the calculation of a local minimizer [5, 13]. The complexity of the QP subproblem has been a major impediment to the formulation of second-derivative SQP methods (although methods based on indefinite QP subproblems have been proposed [7, 8]).
(ii) If $H(x_k, y_k)$ is not positive definite, then $p_k$ may not be a descent direction for the merit function. This implies that an alternative direction must be found or the line search must allow the merit function to increase on some iterations (see, e.g., Grippo, Lampariello and Lucidi [32–34], Toint [44], and Zhang and Hager [49]).

Over the years, algorithm developers have avoided these difficulties by solving a convex QP subproblem defined with a positive semidefinite quasi-Newton approximate Hessian. Some methods follow the convex QP solve with an EQP phase that uses exact second derivatives (see, e.g., [2, 3, 9, 27–29, 37]). However, the common feature of all these approaches is to rely on the convex QP to identify the active constraints at a solution of (NP). In this form, SQP methods have proved reliable and efficient for many problems. For example, under mild conditions the general-purpose solvers `NLPQL` [41], `NPSOL` [20, 22], `DONLP` [43], and `SNOPT7` [23] typically find a (local) optimum from an arbitrary starting point, and they require relatively few evaluations of the problem functions and gradients.

In the next three sections we outline the basic components of an SQP method that incorporates exact second derivatives but avoids the difficulties discussed above.

## 4.2 *Overview of convexification methods*

Convexification is a process for defining a local convex approximation of a nonconvex problem. This approximation may be defined on the full space of variables or just on some subset. Many model-based optimization methods use some form of convexification. For example, line-search methods for unconstrained and linearly constrained optimization define a convex local quadratic model in which the Hessian $H(x_k, y_k)$ is replaced by a positive-definite matrix $H(x_k, y_k) + E_k$ (see, e.g., Greenstadt [31], Gill and Murray [15], Schnabel and Eskow [42], and Forsgren and Murray [12]). All of these methods are based on convexifying an unconstrained or equality-constrained local model. Here we consider a method that convexifies the inequality-constrained subproblem directly. The method extends some approaches proposed by Gill and Robinson [16, Section 4] and Kungurtsev [36].

In the context of SQP methods, the purpose of the convexification is to find a matrix $\Delta H_k$ such that

$$p_k^T\big(H(x_k,y_k)+\Delta H_k\big)p_k \geq \bar{\gamma} p_k^T p_k,$$

for a given primal-dual pair $(x_k, y_k)$, where $\bar{\gamma}$ is a fixed positive scalar that defines a minimum acceptable value of the curvature of the Lagrangian. Ideally, any algorithm for computing $\Delta H_k$ should satisfy two requirements. First, the convexification should be minimal, i.e., if $H(x_k, y_k)$ is positive definite or $p_k^T H(x_k,y_k)p_k \geq \bar{\gamma} p_k^T p_k$, then $\Delta H_k$ should be zero. Second, it must be possible to store the modification $\Delta H_k$ implicitly, without the need to modify the elements of $H(x_k, y_k)$.

The proposed convexification scheme can take three forms: preconvexification, concurrent convexification, and post-convexification. We emphasize that not all of these modifications are necessary at a given iteration.

## 4.3 *Concurrent QP convexification*

Concurrent convexification is based on a specific method for solving a general (i.e., potentially nonconvex) QP (see Gill and Wong [19]). We start by giving a brief description of this method applied to a generic QP of the form

$$\begin{aligned} &\underset{x\in\mathbb{R}^n}{\text{minimize}} \quad \varphi(x) = g^T(x-x_I) + \tfrac{1}{2}(x-x_I)^T H(x-x_I) \\ &\text{subject to} \quad Ax \geq Ax_I - b, \end{aligned} \tag{12}$$

where $x_I$, $b$, $A$, $g$, and $H$ are constant. In the SQP context, $x_I = x_k$, $g = g(x_k)$, $b = c(x_k)$, $A = J(x_k)$, and $H$ is the exact or approximate Hessian of the Lagrangian. Thus, the objective is not necessarily convex and the QP subproblem may be indefinite.

The method described by Gill and Wong in [19] and implemented in the software package `SQIC` [18] is a two-phase method for general QP. In the first phase, the objective function is ignored while a conventional phase-one linear program is used to find a feasible point $x_0$ for the constraints $Ax \geq Ax_I - b$. On completion of the first phase, a working set $\mathcal{W}_0$ is available that contains the indices of a linearly independent subset of the gradients of the active constraints at $x_0$. If $A_0$ and $b_0$ denote the $m_0 \times n$ matrix of rows of $A$ and the vector of $m_0$ components of $b$ with indices in $\mathcal{W}_0$, then

$$A_0 x_0 = A_0 x_I - b_0.$$

In the second phase, a sequence of primal-dual iterates $\{(x_j, y_j)\}_{j\geq 0}$ and working sets $\{\mathcal{W}_j\}$ is generated such that: (i) $\{x_j\}_{j\geq 0}$ is feasible; (ii) $\varphi(x_j) \leq \varphi(x_{j-1})$; and (iii) for every $j \geq 1$, $(x_j, y_j)$ is the primal and dual solution of the equality constrained problem defined by minimizing $\varphi(x)$ subject to the constraints in the working set

$\mathcal{W}_j$. The vector $x_j$ associated with the primal-dual pair $(x_j, y_j)$ is known as a *subspace minimizer* with respect to $\mathcal{W}_j$. If $A_j$ denotes the $m_j \times n$ matrix of rows of $A$ with indices in $\mathcal{W}_j$, then a subspace minimizer is formally defined as the point $x_j$ such that $g(x_j) = A_j^T y_j$, and the KKT matrix

$$K_j = \begin{pmatrix} H & A_j^T \\ A_j & 0 \end{pmatrix} \tag{13}$$

has exactly $m_j$ negative eigenvalues. Equivalently, the associated reduced Hessian $Z_j^T H Z_j$, where the columns of $Z_j$ form a basis for the null-space of $A_j$, is positive definite. Thus, for any $K_j$ satisfying this property, the working set $\mathcal{W}_j$ is *second-order consistent with respect to H*.

In general, the first iterate $x_0$ will not minimize $\varphi(x)$ on $\mathcal{W}_0$, and one or more preliminary iterations are needed to find the first subspace minimizer $x_1$. An estimate of $x_1$ is defined by solving the equality-constrained QP subproblem

$$\underset{x}{\text{minimize}} \;\; \varphi(x) \quad \text{subject to} \;\; A_0(x - x_I) + b_0 = 0. \tag{14}$$

If the KKT matrix $K_0$ is second-order consistent, then the solution of this subproblem is given by $x_0 + p_0$, where $p_0$ satisfies the nonsingular system

$$\begin{pmatrix} H & A_0^T \\ A_0 & 0 \end{pmatrix} \begin{pmatrix} p_0 \\ -\hat{y}_0 \end{pmatrix} = -\begin{pmatrix} g(x_0) \\ b_0 + A_0(x_0 - x_I) \end{pmatrix} = -\begin{pmatrix} g(x_0) \\ 0 \end{pmatrix}. \tag{15}$$

If $x_0 + p_0$ is feasible for (14), then $(x_1, y_1) = (x_0 + p_0, \hat{y}_0)$ is a subspace minimizer; otherwise one of the constraints violated at $x_0 + p_0$ is added to the working set and (15) is solved again with the new working set. Eventually, the working set will include enough constraints to define a primal-dual pair $(x_1, y_1)$ at a subspace minimizer.

If the first subspace minimizer $x_1$ is not optimal for (12), then the method proceeds to find the sequence of subspace minimizers $x_2, x_3, \ldots,$ described above. At any given iteration, not all the constraints in $\mathcal{W}_j$ are necessarily active at $x_j$. If every working-set constraint *is* active, then $\mathcal{W}_j \subseteq \mathcal{A}(x_j)$, and $x_j$ is called a *standard* subspace minimizer; otherwise $x_j$ is a *nonstandard* subspace minimizer. The method is formulated so that there is a subsequence of "standard" iterates intermixed with a finite number of consecutive "nonstandard" iterates. If the multipliers $y_j$ are nonnegative at a standard iterate, then $x_j$ is optimal for (12) and the algorithm is terminated. Otherwise, a working set constraint with a negative multiplier is identified and designated as the *nonbinding working-set constraint* associated with the subsequent consecutive sequence of nonstandard iterates. If the index of the nonbinding constraint corresponds to row $s$ of $A$, then $[y_j]_s < 0$. There follows a sequence of "intermediate" iterations in which the constraint $a_s^T x \geq a_s^T x_I - b_s$ remains in the working set, though it is no longer active, while its multiplier is driven to zero. At each of these iterations, a search direction is defined by solving the equality-constrained subproblem

$$\underset{p\in\mathbb{R}^n}{\text{minimize}}\ \ \varphi(x_j+p) \quad \text{subject to} \ \ a_i^Tp = \begin{cases} 0 & \text{if } i \neq s,\, i\in\mathcal{W}_j, \\ 1 & \text{if } i = s. \end{cases} \tag{16}$$

In matrix form, the optimality conditions for subproblem (16) are

$$\begin{pmatrix} H & A_j^T \\ A_j & 0 \end{pmatrix} \begin{pmatrix} p_j \\ -q_j \end{pmatrix} = \begin{pmatrix} 0 \\ e_s \end{pmatrix}, \tag{17}$$

where $y_j + q_j$ are the multipliers at the minimizer $x_j + p_j$, and $e_s$ denotes the $s$th column of the identity matrix. (To simplify the notation, it is assumed that the nonbinding constraint corresponds to the $s$th row of $A$, which implies that $a_s^T$ is the $s$th row of both $A$ and $A_j$.) Any nonzero step along $p_j$ increases the residual of the nonbinding constraint while maintaining the residuals of the other working-set constraints at zero (i.e., the nonbinding constraint becomes inactive while the other working-set constraints remain active).

Once the direction $(p_j, q_j)$ has been computed, the computation of the next iterate $x_{j+1}$ depends on the value of $p_j^THp_j$, the curvature of $\varphi$ along $p_j$. There are two cases to consider.

**Case 1: $p_j^THp_j > 0$.** In this case the curvature is positive along $p_j$. This will always be the outcome when $\varphi$ is convex. In this case, the step to the minimizer of $\varphi$ along the search direction $p_j$ is given by

$$\alpha_j^* = -g(x_j)^Tp_j/p_j^THp_j = -[y_j]_s/p_j^THp_j. \tag{18}$$

The definition of $\alpha_j^*$ implies that the multiplier $[y_j + \alpha_j^* q_j]_s$ associated with the nonbinding constraint at $x_j + \alpha_j^* p_j$ is zero. This implies that if $x_j + \alpha_j^* p_j$ is feasible with respect to the constraints that are not in the working set, then the nonbinding constraint index can be removed from $\mathcal{W}_j$ as the conditions for a subspace minimizer continue to hold. This gives a new standard iterate $x_{j+1} = x_j + \alpha_j^* p_j$, with working set $\mathcal{W}_{j+1} = \mathcal{W}_j \setminus \{s\}$. Either $x_{j+1}$ is optimal for the QP or a new nonbinding constraint is identified and the process is repeated by computing a search direction from the system (17) defined at $x_{j+1}$. If $x_j + \alpha_j^* p_j$ is not feasible, then $x_{j+1}$ is defined as $x_j + \alpha_j p_j$, where $\alpha_j$ is the largest step that gives a feasible $x_j + \alpha_j p_j$. The point $x_{j+1}$ must have at least one constraint that is active but not in $\mathcal{W}_j$. If $t$ is the index of this constraint, and $a_t$ and the vectors $\{a_i\}_{i\in\mathcal{W}_j}$ are linearly independent, then $t$ is added to the working set to give $\mathcal{W}_{j+1}$. At the next iteration, a new value of $(p_j, q_j)$ is computed using system (17) defined with $A_{j+1}$. If $a_t$ and $\{a_i\}_{i\in\mathcal{W}_j}$ are linearly dependent, then it is shown in [19] that the working set $\mathcal{W}_{j+1} = \{\mathcal{W}_j \setminus \{s\}\} \cup \{t\}$ defined by replacing the index $t$ with index $s$ defines a linearly independent set of constraint gradients. Moreover, $x_{j+1} = x_j + \alpha_j p_j$ is a subspace minimizer with respect to $\mathcal{W}_{j+1}$.

**Case 2: $p_j^T H p_j \le 0$.** In this case $H$ is not positive definite and $\varphi(x_j + \alpha p_j)$ is unbounded below for positive values of $\alpha$. Either the QP is unbounded, or there exists a constraint index $t$ and a nonnegative step $\hat{\alpha}_j$ such that the constraint residuals satisfy $r_t(x_j + \hat{\alpha}_j p_j) = 0$, $r(x_j + \hat{\alpha}_j p_j) \ge 0$, and $\hat{\alpha}_j$ minimizes $\varphi(x_j + \alpha p_j)$ for all feasible $x_j + \alpha p_j$. In this case, $x_{j+1} = x_j + \hat{\alpha}_j p_j$ and, as above, either $a_t$ and $\{a_i\}_{i \in \mathcal{W}_j}$ are linearly independent, in which case $\mathcal{W}_{j+1} = \mathcal{W}_j \cup \{t\}$, or the constraint gradients associated with the working set defined by replacing the index $t$ with index $s$ are linearly independent. Moreover, $x_{j+1} = x_j + \alpha_j p_j$ is a subspace minimizer with respect to $\mathcal{W}_{j+1}$.

To determine whether $a_t$ and the vectors $\{a_i\}_{i \in \mathcal{W}_j}$ are linearly independent, a second KKT system of the form

$$\begin{pmatrix} H & A_j^T \\ A_j & 0 \end{pmatrix} \begin{pmatrix} u_j \\ -v_j \end{pmatrix} = \begin{pmatrix} a_t \\ 0 \end{pmatrix}$$

is solved. It is shown in [19] that $u_j \neq 0$ if and only if $a_t$ and $\{a_i\}_{i \in \mathcal{W}_j}$ are linearly independent. Furthermore, if $u_j \neq 0$, then $u_j^T a_t > 0$ so that linear independence can be determined by checking the sign of the inner product of $u_j$ and $a_t$.

In both cases, the process is repeated at the next subspace minimizer defined by an appropriate working set until an optimal solution is found or the problem is declared to be unbounded.

The proposed concurrent convexification scheme is based on defining an implicit modification of $H$ when negative curvature is detected following the identification of the nonbinding constraint. Assume that a QP search direction $p_j$ with zero or negative curvature is detected after the selection of $a_s^T x \ge b_s$ as the nonbinding constraint (i.e., $[y_j]_s < 0$). In this case, $H$ is not positive definite and the QP Hessian is modified so that it has sufficiently large positive curvature along $p_j$. As $p_j^T H p_j \le 0$, the objective $\varphi(x_j + \alpha p_j)$ is unbounded below for positive values of $\alpha$. In this case, either the unmodified QP is unbounded, or there exists a constraint index $t$ and a nonnegative step $\hat{\alpha}_j$ such that the constraint residuals satisfy $r_t(x_j + \hat{\alpha}_j p_j) = 0$, $r(x_j + \hat{\alpha}_j p_j) \ge 0$, and $\hat{\alpha}_j$ minimizes $\varphi(x_j + \alpha p_j)$ for all feasible $x_j + \alpha p_j$.

If $p_j^T H p_j < 0$, the positive semidefinite rank-one matrix $\sigma a_s a_s^T$ is added to $H$ implicitly. This modifies the quadratic program being solved, but the current iterate $x_j$ remains a subspace minimizer for the modified problem. The only computed quantities altered by the modification are the curvature and the multiplier $y_s$ associated with the nonbinding working-set constraint. The modified Hessian is defined as $H(\bar{\sigma}) = H + \bar{\sigma} a_s a_s^T$ for some $\bar{\sigma} > 0$. Gill and Wong [19] show that the curvature $p^T H p$ is nondecreasing during a sequence of nonstandard iterations associated with a nonbinding index $s$. This implies that a modification of the Hessian will occur only at the first nonstandard iterate.

For an arbitrary $\sigma$, the gradient of the modified objective at $x_j$ is

$$g + H(\sigma)(x_j - x_I) = g + (H + \sigma a_s a_s^T)(x_j - x_I).$$

As $(x_j, y_j)$ is a standard subspace minimizer for the unmodified problem, the identities $g(x_j) = g + H(x_j - x_I) = A_j^T y_j$ and $a_s^T(x_j - x_I) = -b_s$ hold, and the gradient of the modified objective is given by

$$
\begin{aligned}
g + H(\sigma)(x_j - x_I) &= g + H(x_j - x_I) + \sigma a_s a_s^T(x_j - x_I) \\
&= g(x_j) + \sigma a_s^T(x_j - x_I) a_s \\
&= A_j^T(y_j - \sigma b_s e_s) = A_j^T y(\sigma), \quad \text{with} \quad y(\sigma) = y_j - \sigma b_s e_s.
\end{aligned}
$$

This implies that $x_j$ is a subspace minimizer of the modified problem for all $\sigma \geq 0$. Moreover, the multipliers of the modified problem are the same as those of the unmodified problem except for the multiplier $y_s$ associated with the nonbinding constraint, which is shifted by $-\sigma b_s$.

Once the Hessian is modified, the system (17) for the primal-dual direction becomes

$$
\begin{pmatrix} H + \bar{\sigma} a_s a_s^T & A_j^T \\ A_j & 0 \end{pmatrix} \begin{pmatrix} \bar{p}_j \\ -\bar{q}_j \end{pmatrix} = \begin{pmatrix} 0 \\ e_s \end{pmatrix},
$$

which is equivalent to

$$
\begin{pmatrix} H & A_j^T \\ A_j & 0 \end{pmatrix} \begin{pmatrix} p_j \\ -(\bar{q}_j - \bar{\sigma} e_s) \end{pmatrix} = \begin{pmatrix} 0 \\ e_s \end{pmatrix}.
$$

A comparison with (17) yields

$$
\bar{p}_j = p_j \quad \text{and} \quad \bar{q}_j = q_j + \bar{\sigma} e_s,
$$

which implies that the QP direction is not changed by the modification.

For any $\sigma \geq 0$, let $\alpha_j(\sigma)$ denote the step associated with the search direction for the modified QP. The identities $a_s^T p_j = 1$, $a_s^T(x_j - x_I) = -b_s$ and $y_s = g(x_j)^T p_j$ imply that

$$
\begin{aligned}
\alpha_j(\sigma) &= -\frac{\left(g + (H + \sigma a_s a_s^T)(x_j - x_I)\right)^T p_j}{p_j^T(H + \sigma a_s a_s^T) p_j} \\
&= -\frac{g(x_j)^T p_j + \sigma a_s^T(x_j - x_I)}{p_j^T H p_j + \sigma} \\
&= -\frac{g(x_j)^T p_j - \sigma b_s}{p_j^T H p_j + \sigma} = -\frac{y_s - \sigma b_s}{p_j^T H p_j + \sigma} = -\frac{y_s(\sigma)}{p_j^T H p_j + \sigma}.
\end{aligned}
\tag{19}
$$

This implies that $\bar{\sigma}$ must be chosen large enough to satisfy

$$
\bar{\sigma} > \sigma_{\min} = -p_j^T H p_j.
$$

The derivative of $\alpha_j(\sigma)$ with respect to $\sigma$ is given by

$$\alpha_j'(\sigma) = \frac{1}{(p_j^T H p_j + \sigma)^2}\left(y_s + b_s p_j^T H p_j\right) = \frac{y_s(\sigma_{\min})}{(p_j^T H p_j + \sigma)^2}. \tag{20}$$

The choice of $\bar{\sigma}$ that we propose depends on two parameters $y_{\mathtt{tol}}$ and $d_{\max}$. The scalar $d_{\max}$ defines the maximum change in $x$ at each QP iteration. The scalar $y_{\mathtt{tol}}$ is the dual optimality tolerance and is used to define what is meant by a "nonoptimal" multiplier. In particular, the multiplier of the nonbinding constraint must satisfy $y_s < -y_{\mathtt{tol}}$ in order to qualify as being nonoptimal.

There are two cases to consider for the choice of $\bar{\sigma}$.

**Case (i):** $b_s < 0$. In this case, $y_s(\sigma)$ is an increasing function of $\sigma$, which implies that there exists $\sigma_{\mathtt{opt}} = (y_s - y_{\mathtt{tol}})/b_s > 0$ such that $y_s(\sigma_{\mathtt{opt}}) = y_{\mathtt{tol}} > 0$. This modification changes the multiplier associated with the nonbinding constraint from nonoptimal to optimal. However, if $\sigma_{\mathtt{opt}} < \sigma_{\min}$, then the curvature is not sufficiently positive and $\sigma$ must be increased so that it is larger than $\sigma_{\min}$. The definition

$$\bar{\sigma} = \begin{cases} \sigma_{\mathtt{opt}} & \text{if } \sigma_{\mathtt{opt}} \geq 2\sigma_{\min}; \\ 2\sigma_{\min} & \text{if } \sigma_{\mathtt{opt}} < 2\sigma_{\min}, \end{cases}$$

guarantees that the curvature along $p_j$ is sufficiently positive with an optimal modified multiplier $y_s(\bar{\sigma})$. In either case, the QP algorithm proceeds by selecting an alternative nonbinding constraint without taking a step along $p_j$.

If $y_s(\sigma_{\min}) < 0$, it is possible to choose $\bar{\sigma} > \sigma_{\min}$ such that $y_s(\bar{\sigma})$ remains negative. The multiplier $y_s(\sigma)$ increases from the negative value $y_s(\sigma_{\min})$ to the value $-y_{\mathtt{tol}}$ as $\sigma$ increases from $\sigma_{\min}$ to the positive value $\sigma_{\mathtt{nonopt}} = (y_s + y_{\mathtt{tol}})/b_s$. This implies that if $\sigma$ is chosen in the range $\sigma_{\min} < \sigma \leq \sigma_{\mathtt{nonopt}}$, then the multiplier for the nonbinding constraint remains nonoptimal, and it is possible to both convexify and keep the current nonbinding constraint. However, in the SQP context it is unusual for a nonbinding constraint to have a negative value of $b_s$ when $x_k$ is far from a solution. For an SQP subproblem, $b$ is the vector $c(x_k)$, and a negative value of $b_s$ implies that the $s$th nonlinear constraint is violated at $x_k$. The linearization of a violated nonlinear constraint is likely to be retained in the working set because the SQP step is designed to reduce the nonlinear constraint violations. The picture changes when $x_k$ is close to a solution of (NP) and the violations of the nonlinear constraints in the QP working set are small. In this case, if strict complementarity does not hold at the solution of the nonlinear problem[2] and $x_k$ is converging to a point that satisfies the second-order necessary conditions, but not a second-order sufficient condition, then both $b_s$ and $y_s$ may

[2] i.e., $c_j(x^*)y_j^* = 0$ and $c_j(x^*) + y_j^* > 0$ at the optimal primal-dual pair $(x^*, y^*)$.

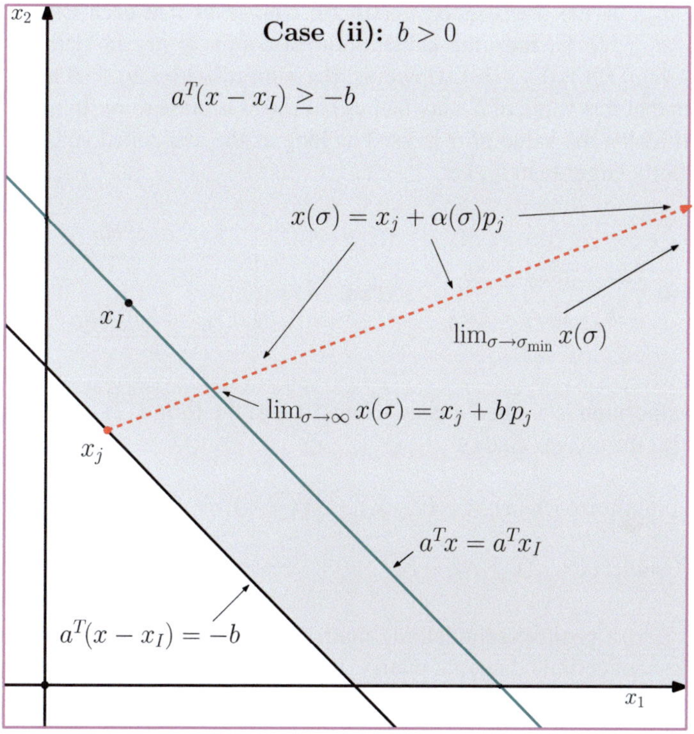


**Fig. 6** The figure depicts the feasible region for a QP with constraints $a^T(x - x_I) \geq -b$, $x_1 \geq 0$, and $x_2 \geq 0$. The point $x_j$ is a standard subspace minimizer with working-set constraint $a^T(x - x_I) \geq -b$. The surface of the hyperplane $a^T(x - x_I) = 0$ is marked in green. The QP base point $x_I$ is feasible for $b \geq 0$. The QP search direction is the red dotted line. The next iterate of the QP algorithm lies on the ray $x(\sigma) = x_j + \alpha_j(\sigma)p_j$. As the modification parameter $\sigma$ increases from its initial value of $\sigma_{\min}$, the new iterate $x(\sigma)$ moves closer to the point $x_j + b\,p_j$ on the hyperplane $a^T(x - x_I) = 0$

be small and negative. It is for this reason that even if $y_s(\sigma_{\min})$ is negative, $\bar{\sigma}$ is chosen large enough that the multiplier changes sign and the nonbinding constraint is retained in the QP working set.

**Case (ii):** $b_s \geq 0$. In this case, $y_s(\sigma_{\min}) = y_s - b_s\sigma_{\min} < 0$ and $y_s(\sigma_{\min})$ decreases monotonically for all increasing $\sigma > \sigma_{\min}$. The step-length function $\alpha_j(\sigma)$ has a pole at $\sigma = -p_j^T H p_j$ and decreases monotonically, with $\alpha_j(\sigma) \to b_s \geq 0$ as $\sigma \to +\infty$. The behavior of $x(\sigma)$ is depicted in Figure 6 for a two-variable QP with constraints $a^T(x - x_I) \geq -b$, $x_1 \geq 0$, and $x_2 \geq 0$. The next iterate of the QP algorithm lies on the ray $x(\sigma) = x_j + \alpha_j(\sigma)p_j$. As $\sigma \to \infty$, $x(\sigma)$ moves closer to the point $x_j + b_s p_j$ on the hyperplane $a^T(x - x_I) = 0$.

A preliminary value of $\bar{\sigma}$ is chosen to give an $x_{j+1}$ such that

$$\|x_{j+1} - x_j\|_2 \leq d_{\max},$$

where $d_{\max}$ is the preassigned maximum change in $x$ at each QP iteration. If $\alpha_T = d_{\max}/\|p_j\|_2$, then the substitution of $\alpha_j(\bar{\sigma}) = \alpha_T$ in (19) gives $\bar{\sigma} = -(y_s + \alpha_T p_j^T H p_j)/(\alpha_T - b_s)$. However, the limit $\alpha_j(\sigma) \to b_s \geq 0$ as $\sigma \to +\infty$ implies that this value of $\bar{\sigma}$ may be large if $\alpha_j(\bar{\sigma})$ is close to $b_s$. In order to avoid this difficulty, the value of $\bar{\sigma}$ is used as long as the associated value of $\alpha_j(\bar{\sigma})$ is sufficiently larger than $b_s$, i.e.,

$$\alpha_j(\bar{\sigma}) = \begin{cases} \alpha_T & \text{if } \alpha_T \geq 2b_s; \\ 2b_s & \text{if } \alpha_T < 2b_s, \end{cases} \quad \text{so that} \quad \bar{\sigma} = \begin{cases} -\dfrac{y_s + \alpha_T p_j^T H p_j}{\alpha_T - b_s} & \text{if } \alpha_T \geq 2b_s, \\[2ex] -\dfrac{y_s + 2b_s p_j^T H p_j}{b_s} & \text{if } \alpha_T < 2b_s. \end{cases}$$

If this algorithm is applied to a nonconvex QP of the form (12), then a solution is found for the convexified QP

$$\begin{aligned} &\underset{x \in \mathbb{R}^n}{\text{minimize}} && \varphi(x) = g^T(x - x_I) + \tfrac{1}{2}(x - x_I)^T(H + E)(x - x_I) \\ &\text{subject to} && Ax \geq Ax_I - b, \end{aligned} \tag{21}$$

where $E$ is a positive-semidefinite matrix of the form $E = A^T \bar{\Sigma} A$, with $\bar{\Sigma}$ a positive semidefinite diagonal matrix. In general, most of the diagonal elements of $\bar{\Sigma}$ are zero. The modification $E$ may be reconstructed from $A$ and a sparse representation of $\bar{\Sigma}$.

## 4.4 *Preconvexification*

The concurrent convexification method of Section 4.3 has the property that if $x_0$ is a subspace minimizer, then all subsequent iterates are subspace minimizers. Methods for finding an initial subspace minimizer utilize an initial estimate $x_0$ of the solution together with an initial working set $\mathcal{W}_0$ of linearly independent constraint gradients. These estimates are often available from a phase-one linear program or, in the SQP context, the solution of the previous QP subproblem.

If the KKT matrix $K_0$ defined by these initial estimates has too many negative or zero eigenvalues, then $\mathcal{W}_0$ is not a second-order consistent working set. In this case, an appropriate $K_0$ may be obtained by imposing temporary constraints that are deleted during the course of the subsequent QP iterations. For example, if $n$ variables are temporarily fixed at their current values, then $A_0$ is the identity matrix and $K_0$ necessarily has exactly $n$ negative eigenvalues regardless of the eigenvalues of $H(x_k, y_k)$. The form of the temporary constraints depends on the method used to solve the KKT equations; see, e.g., Gill and Wong [19, Section 6]. Once the temporary constraints are imposed, concurrent convexification can proceed as in Section 4.3 as the temporary constraints are removed from the working set during subsequent iterations.

A disadvantage of using temporary constraints is that it may be necessary to factor two KKT matrices if the initial working set is not second-order consistent. An alternative approach is to utilize the given working set $\mathcal{W}_0$ without modification and use *preconvexification*, which involves the definition of a positive-semidefinite $E_0$ such that the matrix

$$K_0 = \begin{pmatrix} H + E_0 & A_0^T \\ A_0 & 0 \end{pmatrix} \tag{22}$$

is second-order consistent. A suitable modification $E_0$ may be based on some variant of the symmetric indefinite or block-triangular factorizations of $K_0$. Appropriate methods include: (i) the inertia controlling $\mathrm{LBL}^T$ factorization (Forsgren [10], Forsgren and Gill [11]); (ii) an $\mathrm{LBL}^T$ factorization with pivot modification (Gould [26]); and (iii) a conventional $\mathrm{LBL}^T$ factorization of (22) with $E_0 = \sigma I$ for some nonnegative scalar $\sigma$ (Wächter and Biegler [46]). In each case, the modification $E_0$ is zero if $\mathcal{W}_0$ is already second-order consistent.

## 4.5 *Post-convexification*

As concurrent convexification generates a sequence of second-order-consistent working sets, the SQP search direction $p_k = \hat{x}_k - x_k$ must satisfy the second-order-consistent KKT system

$$\begin{pmatrix} H_k + E_k & J_w(x_k)^T \\ J_w(x_k) & 0 \end{pmatrix} \begin{pmatrix} p_k \\ -\hat{y}_w \end{pmatrix} = -\begin{pmatrix} g(x_k) \\ c_w(x_k) \end{pmatrix}, \tag{23}$$

where $H_k = H(x_k, y_k)$ is the exact Hessian of the Lagrangian, $E_k$ is the matrix defined by the pre- and/or concurrent convexification, and $c_w(x_k)$ and $J_w(x_k)$ are the rows of $c(x_k)$ and $J(x_k)$ associated with indices in the final QP working set $\mathcal{W}$ (cf. (7)). In most cases, concurrent convexification is sufficient to give $p_k^T(H_k + E_k)p_k > 0$, but it may hold that $p_k^T(H_k + E_k)p_k \le 0$. In this case, $p_k$ is not a descent direction for $g_L(x_k, \hat{y}_k)$, and an additional *post-convexification step* is necessary. In the following discussion, there is no loss of generality in assuming that $E_k = 0$, i.e., it is assumed that $H_k$ has not been modified during the preconvexification or concurrent convexification stages. Post-convexification is based on the following result.

**Result 4.1** *If $J_w$ is a second-order-consistent working-set matrix associated with a symmetric $H$, then there exists a nonnegative $\bar{\sigma}$ such that the matrix $\bar{H} = H + \bar{\sigma} J_w^T J_w$ is positive definite. In addition, the solutions of the systems*

$$\begin{pmatrix} H & J_w^T \\ J_w & 0 \end{pmatrix}\begin{pmatrix} p \\ -\hat{y}_w \end{pmatrix} = -\begin{pmatrix} g \\ c_w \end{pmatrix} \quad \textit{and} \quad \begin{pmatrix} \bar{H} & J_w^T \\ J_w & 0 \end{pmatrix}\begin{pmatrix} \bar{p} \\ -\bar{y}_w \end{pmatrix} = -\begin{pmatrix} g \\ c_w \end{pmatrix}$$

*are related by the identities* $\bar{p} = p$ *and* $\bar{y}_w = \hat{y}_w - \bar{\sigma} c_w$. ∎

If the solution $(\hat{x}_k, \hat{y}_k)$ of the QP subproblem does not satisfy the descent condition, then $p_k = \hat{x}_k - x_k$ is such that

$$p_k^T H(x_k, y_k) p_k = -g_L(x_k, \hat{y}_k)^T p_k < \bar{\gamma} p_k^T p_k$$

for some positive $\bar{\gamma}$. The result implies that multipliers $\bar{y}_k$ such that $[\bar{y}_k]_i = 0$, for $i \notin \mathscr{W}$, and $[\bar{y}_k]_w = \hat{y}_w - \bar{\sigma} c_w(x_k)$, provide the required curvature

$$p_k^T \bar{H}(x_k, y_k) p_k = -g_L(x_k, \bar{y}_k)^T p_k = \gamma p_k^T p_k,$$

where $\bar{\sigma} = \big(\gamma p_k^T p_k - p_k^T H(x_k, y_k) p_k\big) / \|c_w(x_k)\|^2$ with $\gamma$ chosen such that $\gamma \geq \bar{\gamma}$. (If $c_w(x_k) = 0$, then $p_k$ is a descent direction for the objective function and no post-convexification is required; see, e.g., Gill, Murray, Saunders, and Wright [22].) The extension of this result to the situation where $(\hat{x}_k, \hat{y}_k)$ satisfies the modified KKT system (23) is obvious.

## 5 Summary

The numerical results presented in Section 3 indicate that the optimization packages SNOPT7 and IPOPT are very efficient and robust in general, solving over 85% of problems in the CUTEst test collection. However, the results also show that the performance of these codes depends greatly on the characteristics of the problem. These characteristics include the size of the problem, the availability of first and second derivatives, the types of constraints, and the availability of a good initial starting point. Ultimately, for every problem that is best solved by an SQP code, there will likely exist another that is best solved by an IP code.

To extend SQP methods so that second derivatives may be exploited reliably and efficiently, we propose convexification algorithms for the QP subproblem in an active-set SQP method for nonlinearly constrained optimization. Three forms of convexification are defined: preconvexification, concurrent convexification, and post-convexification. The methods require only minor changes to the algorithms used to solve the QP subproblem, and are designed so that modifications to the original problem are minimized and applied only when necessary.

It should be noted that the post-convexification Result 4.1 holds even if a conventional general QP method is used to solve the QP subproblem (provided that the method gives a final working set that is second-order consistent). It follows that post-convexification will define a descent direction regardless of whether concurrent convexification is used or not. The purpose of concurrent convexification is to reduce the probability of needing post-convexification, and to avoid the difficulties associated with solving an indefinite QP problem.

The methods defined here are the basis of the second-derivative solvers in the dense SQP package `DNOPT` of Gill, Saunders, and Wong [25] and the forthcoming `SNOPT9`. All of the methods may be extended to problems in which the constraints are written in the form (9) (see Gill and Wong [19, Section 4]). In this case, the inequality constraints for the QP subproblem are upper and lower bounds, and all the modification matrices are diagonal.

**Acknowledgements** We would like to thank Nick Gould for providing the latest version of the `CUTEst` test collection. We are also grateful to the referees for constructive comments that resulted in significant improvements in the final manuscript.

The research of the author "Philip E. Gill" was supported in part by National Science Foundation grants DMS-1318480 and DMS-1361421. The research of the author "Michael A. Saunders" was supported in part by the National Institute of General Medical Sciences of the National Institutes of Health [award U01GM102098]. The research of the author "Elizabeth Wong" was supported in part by Northrop Grumman Aerospace Systems. The content is solely the responsibility of the authors and does not necessarily represent the official views of the funding agencies.

## References

1. Byrd, R.H., Lu, P., Nocedal, J., Zhu, C.: A limited memory algorithm for bound constrained optimization. SIAM J. Sci. Comput. **16**, 1190–1208 (1995)
2. Byrd, R.H., Gould, N.I.M., Nocedal, J., Waltz, R.A.: An algorithm for nonlinear optimization using linear programming and equality constrained subproblems. Math. Program. **100**(1, Ser. B), 27–48 (2004)
3. Byrd, R.H., Gould, N.I.M., Nocedal, J., Waltz, R.A.: On the convergence of successive linear-quadratic programming algorithms. SIAM J. Optim. **16**(2), 471–489 (2005)
4. Byrd, R., Nocedal, J., Waltz, R., Wu, Y.: On the use of piecewise linear models in nonlinear programming. Math. Program. **137**, 289–324 (2013)
5. Contesse, L.B.: Une caractérisation complète des minima locaux en programmation quadratique. Numer. Math. **34**, 315–332 (1980)
6. Dolan, E.D., Moré, J.J.: Benchmarking optimization software with COPS. Technical Memorandum ANL/MCS-TM-246, Argonne National Laboratory, Argonne (2000)
7. Fletcher, R.: An $\ell_1$ penalty method for nonlinear constraints. In: Boggs, P.T., Byrd, R.H., Schnabel, R.B. (eds.) Numerical Optimization 1984, pp. 26–40. SIAM, Philadelphia (1985)
8. Fletcher, R., Leyffer, S.: User manual for filterSQP. Tech. Rep. NA/181, Dept. of Mathematics, University of Dundee, Scotland (1998)
9. Fletcher, R., Sainz de la Maza, E.: Nonlinear programming and nonsmooth optimization by successive linear programming. Math. Program. **43**, 235–256 (1989)
10. Forsgren, A.: Inertia-controlling factorizations for optimization algorithms. Appl. Num. Math. **43**, 91–107 (2002)
11. Forsgren, A., Gill, P.E.: Primal-dual interior methods for nonconvex nonlinear programming. SIAM J. Optim. **8**, 1132–1152 (1998)
12. Forsgren, A., Murray, W.: Newton methods for large-scale linear equality-constrained minimization. SIAM J. Matrix Anal. Appl. **14**, 560–587 (1993)

13. Forsgren, A., Gill, P.E., Murray, W.: On the identification of local minimizers in inertia-controlling methods for quadratic programming. SIAM J. Matrix Anal. Appl. **12**, 730–746 (1991)
14. Gill, P.E., Leonard, M.W.: Limited-memory reduced-Hessian methods for large-scale unconstrained optimization. SIAM J. Optim. **14**, 380–401 (2003)
15. Gill, P.E., Murray, W.: Newton-type methods for unconstrained and linearly constrained optimization. Math. Program. **7**, 311–350 (1974)
16. Gill, P.E., Robinson, D.P.: A globally convergent stabilized SQP method. SIAM J. Optim. **23**(4), 1983–2010 (2013)
17. Gill, P.E., Wong, E.: Sequential quadratic programming methods. In: Lee, J., Leyffer, S. (eds.) Mixed Integer Nonlinear Programming. The IMA Volumes in Mathematics and its Applications, vol. 154, pp. 147–224. Springer New York (2012). http://dx.doi.org/10.1007/978-1-4614-1927-3_6. 10.1007/978-1-4614-1927-3_6
18. Gill, P.E., Wong, E.: User's guide for SQIC: software for large-scale quadratic programming. Center for Computational Mathematics Report CCoM 14-02, Center for Computational Mathematics, University of California, San Diego, La Jolla (2014)
19. Gill, P.E., Wong, E.: Methods for convex and general quadratic programming. Math. Program. Comput. **7**, 71–112 (2015). doi:10.1007/s12532-014-0075-x. http://dx.doi.org/10.1007/s12532-014-0075-x
20. Gill, P.E., Murray, W., Saunders, M.A., Wright, M.H.: User's guide for NPSOL (Version 4.0): a Fortran package for nonlinear programming. Report SOL 86-2, Department of Operations Research, Stanford University, Stanford (1986)
21. Gill, P.E., Murray, W., Saunders, M.A., Wright, M.H.: Maintaining *LU* factors of a general sparse matrix. Linear Algebra Appl. **88/89**, 239–270 (1987). doi:10.1016/0024-3795(87)90112-1. http://dx.doi.org/10.1016/0024-3795(87)90112-1
22. Gill, P.E., Murray, W., Saunders, M.A., Wright, M.H.: Some theoretical properties of an augmented Lagrangian merit function. In: Pardalos, P.M. (ed.) Advances in Optimization and Parallel Computing, pp. 101–128. North Holland, Amsterdam (1992)
23. Gill, P.E., Murray, W., Saunders, M.A.: SNOPT: an SQP algorithm for large-scale constrained optimization. SIAM Rev. **47**, 99–131 (2005)
24. Gill, P.E., Murray, W., Saunders, M.A.: User's guide for SQOPT Version 7: software for large-scale linear and quadratic programming. Numerical Analysis Report 06-1, Department of Mathematics, University of California, San Diego, La Jolla (2006)
25. Gill, P.E., Saunders, M.A., Wong, E.: User's Guide for DNOPT: a Fortran package for medium-scale nonlinear programming. Center for Computational Mathematics Report CCoM 14-05, Department of Mathematics, University of California, San Diego, La Jolla (2014)
26. Gould, N.I.M.: On modified factorizations for large-scale linearly constrained optimization. SIAM J. Optim. **9**, 1041–1063 (1999)
27. Gould, N.I.M., Robinson, D.P.: A second derivative SQP method with imposed descent. Numerical Analysis Report 08/09, Computational Laboratory, University of Oxford, Oxford (2008)
28. Gould, N.I.M., Robinson, D.P.: A second derivative SQP method: global convergence. SIAM J. Optim. **20**(4), 2023–2048 (2010)
29. Gould, N.I.M., Robinson, D.P.: A second derivative SQP method: local convergence and practical issues. SIAM J. Optim. **20**(4), 2049–2079 (2010)
30. Gould, N.I.M., Orban, D., Toint, P.L.: CUTEst: a constrained and unconstrained testing environment with safe threads. Technical report, Rutherford Appleton Laboratory, Chilton (2013). doi:10.1007/s10589-014-9687-3. http://dx.doi.org/10.1007/s10589-014-9687-3
31. Greenstadt, J.: On the relative efficiencies of gradient methods. Math. Comput. **21**, 360–367 (1967)
32. Grippo, L., Lampariello, F., Lucidi, S.: Newton-type algorithms with nonmonotone line search for large-scale unconstrained optimization. In: System modelling and optimization (Tokyo, 1987). Lecture Notes in Control and Inform. Sci., vol. 113, pp. 187–196. Springer, Berlin (1988). doi:10.1007/BFb0042786. http://dx.doi.org/10.1007/BFb0042786

33. Grippo, L., Lampariello, F., Lucidi, S.: A truncated Newton method with nonmonotone line search for unconstrained optimization. J. Optim. Theory Appl. **60**(3), 401–419 (1989). doi:10.1007/BF00940345. http://dx.doi.org/10.1007/BF00940345
34. Grippo, L., Lampariello, F., Lucidi, S.: A class of nonmonotone stabilization methods in unconstrained optimization. Numer. Math. **59**(8), 779–805 (1991). doi:10.1007/BF01385810. http://dx.doi.org/10.1007/BF01385810
35. Han, S.P.: A globally convergent method for nonlinear programming. J. Optim. Theory Appl. **22**, 297–309 (1977)
36. Kungurtsev, V.: Second-derivative sequential quadratic programming methods for nonlinear optimization. Ph.D. thesis, Department of Mathematics, University of California San Diego, La Jolla (2013)
37. Morales, J.L., Nocedal, J., Wu, Y.: A sequential quadratic programming algorithm with an additional equality constrained phase. IMA J. Numer. Anal. **32**, 553–579 (2012)
38. Paige, C.C., Saunders, M.A.: Solution of sparse indefinite systems of linear equations. SIAM J. Numer. Anal. **12**, 617–629 (1975)
39. Powell, M.J.D.: A fast algorithm for nonlinearly constrained optimization calculations. In: Watson, G.A. (ed.) Numerical Analysis, Dundee 1977, no. 630 in Lecture Notes in Mathematics, pp. 144–157. Springer, Heidelberg, Berlin, New York (1978)
40. Schittkowski, K.: The nonlinear programming method of Wilson, Han, and Powell with an augmented Lagrangian type line search function. I. Convergence analysis. Numer. Math. **38**(1), 83–114 (1981/1982). doi:10.1007/BF01395810. http://dx.doi.org/10.1007/BF01395810
41. Schittkowski, K.: NLPQL: a Fortran subroutine for solving constrained nonlinear programming problems. Ann. Oper. Res. **11**, 485–500 (1985/1986)
42. Schnabel, R.B., Eskow, E.: A new modified Cholesky factorization. SIAM J. Sci. Stat. Comput. **11**, 1136–1158 (1990)
43. Spellucci, P.: An SQP method for general nonlinear programs using only equality constrained subproblems. Math. Program. **82**, 413–448 (1998)
44. Toint, P.L.: An assessment of nonmonotone linesearch techniques for unconstrained optimization. SIAM J. Sci. Comput. **17**(3), 725–739 (1996). doi:10.1137/S106482759427021X. http://dx.doi.org/10.1137/S106482759427021X
45. Wächter, A., Biegler, L.T.: Line search filter methods for nonlinear programming: local convergence. SIAM J. Optim. **16**(1), 32–48 (electronic) (2005). doi:10.1137/S1052623403426544. http://dx.doi.org/10.1137/S1052623403426544
46. Wächter, A., Biegler, L.T.: Line search filter methods for nonlinear programming: motivation and global convergence. SIAM J. Optim. **16**(1), 1–31 (electronic) (2005). doi:10.1137/S1052623403426556. http://dx.doi.org/10.1137/S1052623403426556
47. Wächter, A., Biegler, L.T.: On the implementation of an interior-point filter line-search algorithm for large-scale nonlinear programming. Math. Program. **106**(1, Ser. A), 25–57 (2006)
48. Wächter, A., Biegler, L.T., Lang, Y.D., Raghunathan, A.: IPOPT: an interior point algorithm for large-scale nonlinear optimization (2002). https://projects.coin-or.org/Ipopt
49. Zhang, H., Hager, W.W.: A nonmonotone line search technique and its application to unconstrained optimization. SIAM J. Optim. **14**(4), 1043–1056 (electronic) (2004). doi:10.1137/S1052623403428208. http://dx.doi.org/10.1137/S1052623403428208

MIX
Papier aus verantwortungsvollen Quellen
Paper from responsible sources
FSC® C105338
FSC
www.fsc.org

If you have any concerns about our products,
you can contact us on
**ProductSafety@springernature.com**

In case Publisher is established outside the EU,
the EU authorized representative is:
**Springer Nature Customer Service Center GmbH**
**Europaplatz 3, 69115 Heidelberg, Germany**

Printed by Libri Plureos GmbH
in Hamburg, Germany